# PALAEOBIOLOGY OF CONODONTS

# BRITISH MICROPALAEONTOLOGICAL SOCIETY SERIES

This series, published by Ellis Horwood Limited for the British Micropalaeontological Society, aims to gather together knowledge of a particular faunal group for specialist and non-specialist geologists alike. The original series of Stratigraphic Atlas or Index volumes ultimately will cover all groups and will describe and illustrate the common elements of the microfauna through time (whether index or long-ranging species) thus enabling the reader to identify characteristic species in addition to those of restricted stratigraphic range. The series has now been enlarged to include the reports of conferences, organized by the Society, and collected essays on specialist themes.

The synthesis of knowledge presented in the series will reveal its strengths and prove its usefulness to the practising micropalaeontologist, and to those teaching and learning the subject. By identifying some of the gaps in this knowledge, the series will, it is believed, promote and stimulate further active research and investigation.

**STRATIGRAPHICAL ATLAS OF FOSSIL FORAMINIFERA**
Editors: D. G. JENKINS, The Open University, and J. W. MURRAY, Professor of Geology, University of Exeter
**MICROFOSSILS FROM RECENT AND FOSSIL SHELF SEAS**
Editors: J. W. NEALE, Professor of Micropalaeontology, University of Hull, and M. D. BRASIER, Lecturer in Geology, University of Hull
**FOSSIL AND RECENT OSTRACODS**
Editors: R. H. BATE, Stratigraphic Services International, Guildford, E. ROBINSON, Department of Geology, University College London, and L. SHEPPARD, Stratigraphic Services International, Guildford
**A STRATIGRAPHICAL INDEX OF CALCAREOUS NANNOFOSSILS**
Editor: A. R. LORD, Department of Geology, University College London
**A STRATIGRAPHICAL INDEX OF CONODONTS**
Editors: A. C. HIGGINS, Geological Survey of Canada, Calgary, and R. L. AUSTIN, Department of Geology, University of Southampton
**CONODONTS: Investigative Techniques and Applications**
Editor: R. L. AUSTIN, Department of Geology, University of Southampton
**PALAEOBIOLOGY OF CONODONTS**
Editor: R. J. ALDRIDGE, Department of Geology, University of Nottingham
**MICROPALAEONTOLOGY OF CARBONATE ENVIRONMENTS**
Editor: M. B. HART, Professor of Micropalaeontology and Head of Department of Geological Studies, Plymouth Polytechnic

# ELLIS HORWOOD SERIES IN GEOLOGY

*Editors:* D. T. DONOVAN, Professor of Geology, University College London, and J. W. MURRAY, Professor of Geology, University of Exeter

This series aims to build up a library of books on geology which will include student texts and also more advanced works of interest to professional geologists and to industry. The series will include translation of important books recently published in Europe, and also books specially commissioned.

**A GUIDE TO CLASSIFICATION IN GEOLOGY**
J. W. MURRAY, Professor of Geology, University of Exeter
**THE CENOZOIC ERA: Tertiary and Quaternary**
C. POMEROL, Professor, University of Paris VI
Translated by D. W. HUMPHRIES, Department of Geology, University of Sheffield, and E. E. HUMPHRIES
Edited by Professor D. CURRY and D. T. DONOVAN, University College London
**INTRODUCTION TO PALAEOBIOLOGY: GENERAL PALAEONTOLOGY**
B. ZIEGLER, Professor of Geology and Palaeontology, University of Stuttgart, and Director of the State Museum for Natural Science, Stuttgart
**FAULT AND FOLD TECTONICS**
W. JAROSZEWSKI, Faculty of Geology, University of Warsaw
**RADIOACTIVITY IN GEOLOGY: Principles and Applications**
E. M. DURRANCE, Department of Geology, University of Exeter

# ELLIS HORWOOD SERIES IN APPLIED GEOLOGY

The books listed below are motivated by the up-to-date applications of geology to a wide range of industrial and environmental factors: they are practical, for use by the professional and practising geologist or engineer, for use in the field, for study, and for reference.

**A GUIDE TO PUMPING TESTS**
F. C. BRASSINGTON, Principal Hydrogeologist, North West Water Authority
**QUATERNARY GEOLOGY: Processes and Products**
JOHN A. CATT, Rothamsted Experimental Station, Harpenden, UK
**PRACTICAL PEDOLOGY: Manual of Soil Formation, Description and Mapping**
S. G. McRAE and C. P. BURNHAM, Department of Environmental Studies and Countryside Planning, Wye College (University of London)

# PALAEOBIOLOGY OF CONODONTS

*Editor:*

RICHARD J. ALDRIDGE, B.Sc.(Hons.), Ph.D.
Reader in Palaeontology
Department of Geology, University of Nottingham

**ELLIS HORWOOD LIMITED**
Publishers · Chichester

for
**THE BRITISH MICROPALAEONTOLOGICAL SOCIETY**

First published in 1987
**ELLIS HORWOOD LIMITED**
Market Cross House, Cooper Street,
Chichester, West Sussex, PO19 1EB, England
*The publisher's colophon is reproduced from James
Gillison's drawing of the ancient Market Cross, Chichester.*


**Distributors:**
*Australia and New Zealand:*
JACARANDA WILEY LIMITED
GPO Box 859, Brisbane, Queensland 4001,
Australia
*Canada:*
JOHN WILEY & SONS CANADA LIMITED
22 Worcester Road, Rexdale, Ontario, Canada
*Europe and Africa:*
JOHN WILEY & SONS LIMITED
Baffins Lane, Chichester, West Sussex, England
*North and South America and the rest of the world:*
Halsted Press: a division of
JOHN WILEY & SONS
605 Third Avenue, New York, NY 10158, USA

© **1987 British Micropalaeontological Society/
Ellis Horwood Limited**

**British Library Cataloguing in Publication Data**
Palaeobiology of conodonts. —
(British Micropalaeontological Society series)
1. Conodonts
I. Aldridge, Richard J.
II. British Micropalaeontological Society
III. Series 562′2    QE899
**Library of Congress Card No.** 86–21414

ISBN 0–85312–906–1 (Ellis Horwood Limited)
ISBN 0–470–20788–4 (Halsted Press)

Printed in Great Britain by
Butler & Tanner, Frome, Somerset

# Contents

# Contributors

Richard J. Aldridge,
Department of Geology, University of Nottingham, University Park, Nottingham NG7 2RD, England.

Derek E. G. Briggs,
Department of Geology, University of Bristol, Wills Memorial Building, Queens Road, Bristol BS8 1RJ, England.

David L. Clark,
Department of Geology and Geophysics, University of Wisconsin-Madison, Madison, Wisconsin 53706, USA.

Kenneth L. Elliott,
Department of Geology, University of Missouri-Columbia, Columbia, Missouri 65211, USA.

Kevin M. Engel,
Department of Geology, University of Missouri-Columbia, Columbia, Missouri 65211, USA.

Raymond L. Ethington,
Department of Geology, University of Missouri-Columbia, Columbia, Missouri 65211, USA.

Lars E. Fåhræus,
Department of Earth Sciences, Memorial University, St John's, Newfoundland, A1B 3X5, Canada.

Goverdina E. Fåhræus-van Ree,
Thyroid Research Laboratory, Health Sciences Centre, Memorial University, St John's, Newfoundland, A1B 3V6, Canada.

Lennart Jeppsson,
Department of Historical Geology and Palaeontology, University of Lund, Solvegatan 13, S-223 62 Lund, Sweden.

H. Richard Lane,
Research Center, Amoco Production Company, P.O. Box 3385, Tulsa, Oklahoma 74102, USA.

Robert S. Nicoll,
Bureau of Mineral Resources, P.O. Box 378, Canberra City, A.C.T., Australia 2601.

Rodney D. Norby,
Illinois State Geological Survey, Natural Resources Building, 615 E. Peabody Drive, Champaign, Illinois 61820, USA.

Carl B. Rexroad,
Indiana Geological Survey, 611 North Walnut Grove, Bloomington, Indiana 47405, USA.

M. Paul Smith,
Department of Geology, University of Nottingham, University Park, Nottingham NG7 2RD, England; now at Department of Earth Sciences, University of Cambridge, Downing Street, Cambridge CB2 3EQ, England.

Hubert Szaniawski,
Zakład Paleobiologii, Polska Akademia Nauk, Al. Żwirki i Wigury 93, 02–089 Warsaw, Poland.

Willi Ziegler,
Forschungsinstitut Senckenberg, Senckenberganlage 25, D-6000 Frankfurt am Main 1, Federal Republic of Germany.

# Preface

This volume is one of a pair that has arisen largely from papers presented at the Fourth European Conodont Symposium (ECOS IV), held at the University of Nottingham during the period 25–29 July, 1985. The companion volume, on investigative techniques and applications, has been edited by R. L. Austin, and a third, a stratigraphical index of conodonts edited by A. C. Higgins and R. L. Austin, was produced to coincide with the Symposium. These three books together provide a coverage of all the main fields of conodont research, and give a clear indication of the direction of future endeavours. These are very exciting times for the conodont worker, stimulated by the recent discoveries of soft-bodied conodont fossils and by the rapidly increasing recognition of the value of the group in biostratigraphy, palaeoecology, evolutionary studies, thermal maturation evaluation, and the investigation of oceanic geochemistry. It is among the purposes of these volumes to convey that excitement and to reflect the diversity of current research activity.

The European Conodont Symposia have now become established events. The first was held in Marburg, Federal Republic of Germany, in 1971, followed by meetings in the dual centres of Vienna and Prague (1980) and in Lund, Sweden (1982). As has been stressed previously, all of these gatherings have been European only in terms of their location, and all have attracted participation from throughout the international community. The ECOS IV meeting (with two field excursions to the Welsh Borders and S.W. England) was attended by more than 100 conodont specialists from 29 countries, and lectures and posters were presented by speakers of 19 different nationalities. The meeting catered for all aspects of conodont research, but two one-day sessions were reserved for the two major themes: (1) investigative techniques and applications, and (2) conodont palaeobiology. Apart from the introductory chapter, all the contributions in this book were formally presented during the latter thematic session.

The papers which comprise the chapters of this volume mostly fall neatly under one of two headings: (1) the conodont animal and its mineralized apparatus, and (2) patterns of conodont evolution. Chapter 1 provides some historical background as an introduction to these two topics, and endeavours to highlight where the most vigorous current debates are centred. In the following five chapters the themes of conodont affinity and of the architecture and function of the conodont apparatus are addressed from widely differing viewpoints. Szaniawski describes the internal structure of the earliest euconodont elements and their putative ancestors, and stresses the similarities to the grasping spines of modern chaetognaths. Aldridge *et al.* and Smith *et al.* prefer to draw analogies between conodont apparatuses and the bilaterally operative lingual apparatus of myxinoids, while Nicoll and Rexroad view the elements as tissue-covered and gathering food in an altogether more passive mode. Each of these authors is drawn towards a particular hypothesis of conodont relationships, with Szaniawski favouring a link with the chaetognaths, Nicoll a closeness to the cephalochordates, and others an affinity with the myxinoids. The nature of the conodont animal has always been a topic of lively debate, and the current controversy is no less intriguing than those of the past.

In a short but potentially revolutionary contribution, Fåhræus and Fåhræus-van Ree (Chapter 7) present the results of their experiments to fix and stain the organic material locked in the mineralized elements of conodonts. Their recognition of cellular material in their preparations opens fascinating possibilities not just of relevance to the question of conodont affinity but to the whole field of palaeobiology. We must hope that their continued work confirms the promise of this preliminary investigation.

The final four chapters relate to the topic of conodont evolution. Cyclicity in extinction/adaptive radiation events has recently been increasingly recognized in the fossil record as a whole and in some particular groups of organisms. It is evident from the papers presented here that conodont evolutionary history shows a comparable pattern. As an example, Ethington *et al.* (Chapter 8) give a detailed account of the faunal changes through an Early Ordovician interval that appears to record some kind of ecological replacement. Subsequent chapters record patterns of conodont faunal turnover from the Silurian through to the end of the Triassic. Jeppsson relates the development of Silurian conodont faunas to changes in oceanic conditions, with the most diverse and specialized associations occurring at time ss of atypical sedimentation worldwide. Ziegler and

Lane recognize a series of cyclic events through the Devonian and early Carboniferous, with extinctions followed by radiative then gradualistic phases. Similar pulses occur into the Permian and Triassic, but diversity in the later stages of conodont history was always lower than in the earlier peaks, and in the final chapter, Clark traces this decline to the final demise of the group in the Late Triassic. The future of studies in this area holds immense promise, especially as it may be possible to tie evolutionary events directly to sea-water geochemistry through chemical analysis of the conodont elements themselves. Examples of recent investigations of this type were presented at ECOS IV and are included in the companion volume, giving some indication of the wide potential of conodont geochemistry.

Although it is very difficult to judge a meeting when one is closely involved in the organization, ECOS IV appears to have been a success. For this, I particularly have to acknowledge the efforts of my co-members on the organizing committee, Ronald Austin and Paul Smith. Many other friends and colleagues made major contributions, especially Jean Angell, Mark Dean, Jean Pearson, Andy Swift, Simon Tull, and Josie Wilkinson. My wife, Alison, organized a programme for accompanying members and solved a multitude of minor problems. To all of these, and many others, my sincere thanks. Thanks are also due to all those who made this volume possible, including the contributors and the referees. The secretarial load was again borne by Jean Angell, Josie Wilkinson provided cartographic expertise, and James Gillison and the staff of Ellis Horwood gave advice and assistance throughout.

ECOS IV was initiated by the Pander Society and was generously supported by the Universities of Nottingham and Southampton, the British Council, the British Micropalaeontological Society, British Petroleum plc, Britoil plc, E. K. Hull Microslide Company, Ellis Horwood Limited, ERICO, the Midland Bank plc, and Stratigraphic Services International.

Richard J. Aldridge

# 1

# Conodont palaeobiology: a historical review

R. J. Aldridge

## ABSTRACT

The history of conodont studies has been marked by vigorous debates concerning the affinities of the organisms and the function of their skeletal elements and apparatuses. Recent discoveries of complete fossil specimens with preserved soft tissue have concentrated the arguments, but opinions are still divided. The closest affinities of the conodonts, however, seem to lie within the chordates. Several workers now favour functional hypotheses that consider the conodont apparatus as a grasping and shredding mechanism, although others prefer to view the elements as supports of ciliated tentacles in a food-gathering sieve or lophophore.

## 1.1 INTRODUCTION

As recently as 1981, in the *Treatise on Invertebrate Paleontology*, K. J. Müller wrote that the nature of conodonts 'is considered by many paleontologists to be one of the most fundamental unanswered questions in systematic paleontology'. The problem of the zoological affinities of these enigmatic extinct fossils has certainly attracted more general attention and inspired a greater diversity of opinions than any other aspect of conodont palaeobiology. Prior to 1983, in the lack of any direct evidence of the soft tissues of the conodont organism, hypotheses ranged from the carefully considered to the bizarre, and taxonomic assignment from the algae (Fahlbusch 1964) to the vertebrates (Pander 1856, and many others). Müller (1981, pp. W78–80) listed 46 publications that appeared in the period from 1856 to 1975, variously suggesting affinities with plants, conulariids, aschelminthes, gnathostomulids, molluscs, annelids, arthropods, tentaculates, chaetognaths, and chordates. These assignments were commonly based on morphological comparisons of structures in extant organisms with conodont elements, although several of the similarities are superficial. Some hypotheses rested on associations of elements with other fossils, although Hinde (1879) and, later, Rhodes (1952, 1954) reported that conodonts may be found associated occasionally but not consistently with almost every type of contemporaneous marine organism. The discovery of fossil specimens with preserved soft tissue (Briggs *et al*. 1983, Mikulic *et al*. 1985a,b, Aldridge *et al*. 1986) has concentrated the debate, but there are still differences in interpretation of the nature of the animals and of the function of the conodont elements and apparatuses. Consideration of the composition and structure of conodont elements and of the architecture of conodont apparatuses continues to be of fundamental importance in these discussions.

## 1.2 COMPOSITION, STRUCTURE, AND FUNCTION OF CONODONT ELEMENTS

Conodont elements are composed of phosphatic lamellae, which in all but the most primitive types were accreted concentrically and centrifugally during the growth of the unit. Conical phosphatic elements first appear in the fossil record close to the Precambrian–Cambrian boundary, and Cambrian forms display a variety of histological structures (Bengtson 1983b). Bengtson (1976) presumed that three of these types were assignable to the conodonts and differentiated them as protoconodonts, paraconodonts, and euconodonts (Fig. 1.1). In the protoconodont elements a thick phosphatic middle layer is bounded externally and internally by thin organic layers; the arrangement of lamellae in the mineralized layer indicates accretion at the inner surface and the base (Bengtson 1976, 1983b, Szaniawski, 1982, 1983). In contrast, paraconodont elements exhibit phosphatic lamellae accreted at the inner and outer surface, but not continuous around the tip (Müller and Nogami 1971, Bengtson 1976). In euconodont, or 'true conodont', elements there is a division of the unit into a hyaline crown and a basal body. The lamellae of the crown are continuous around the upper surface, while

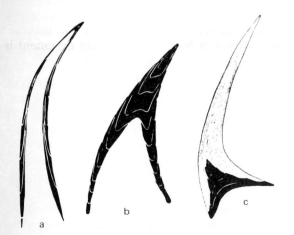

Fig. 1.1—Histology of the phosphatic layers of a) protoconodont, b) paraconodont and c) euconodont elements, after Bengtson (1976), with permission.

those of the basal body may or may not be continuous around the lower surface. The basal body of the euconodont elements may be homologous with the 'cusp' of the paraconodonts (Bengtson 1983b). The vast majority of conodont genera possessed euconodont elements, and the unqualified term 'conodont element' is normally used in reference to euconodonts.

The earliest opinion expressed on the nature of conodont elements was that of their discoverer, C. H. Pander, who in his 1856 monograph regarded them as the teeth of an extinct group of fish. Huddle (1972) recorded that before this publication Pander's specimens had clearly been seen and discussed by several European palaeontologists. In a footnote in his *Siluria*, Murchison (1854, p. 323), reported a letter written by Pander in 1851 to the French Academy of Science referring to minute bodies thought to be fish teeth. Murchison himself felt at that time that they were more likely to be the ends of trilobite segments, and a similar view, that they were crustacean spines, was put forward by Harley (1861). In later editions of *Siluria*, Murchison (1859, p. 374) compared conodont elements to cuttle-fish dentition and reported Richard Owen as concluding that they present 'the most analogy with the spines, hooklets, or denticles of naked Mollusks and Annelides' (Murchison 1867, p. 336). Fuller reports by Owen were given in Appendices to the 1859, 1867, and 1872 editions, and Murchison (1867, p. 336) saw fit to declare that 'the question is completely set at rest'. In reality, the debate had hardly begun.

On the basis of his collections of conodont elements, Pander (1856) reasoned that it was unlikely that other parts of the conodont animal would ever be found. Hinde (1879) agreed that the elements were the only parts of the organism capable of fossilization, and, on the grounds of composition and morphology, rejected the notion that they belonged to annelids or molluscs. He noted similarities with the teeth of existing myxinoids, but recognized that the analogy was imperfect, and judiciously concluded that 'the facts at hand appear insufficient to decide the question' (Hinde 1879, p. 356).

Despite all the uncertainties regarding conodont affinities, the close resemblance of many elements to the teeth of various extant organisms led most workers over the next 75 years to regard them as having been teeth or jaws, with annelid or fish associations particularly in favour (see Globensky 1970). Other functional suggestions during this period included Denham's (1944) theory that they represented the copulatory structures of a group or groups of worms, and the proposals by Huddle (1934) and Hass (1941) that they served as internal supports for soft tissues. In 1954, Rhodes outlined and critically analysed all hypotheses that had been propounded to that date, and refuted Denham's idea on the evidence of morphology, composition, and size of elements. Rhodes' conclusion was that conodont elements were the teeth or jaws of an extinct group, and that only two theories of their affinities deserved further consideration: they belonged either to worm-like creatures or to primitive vertebrates. In his view the composition of elements was all-important, and he posed the question: 'Must the chemical composition of conodonts (calcium phosphate) be interpreted as evidence of a vertebrate origin, or is it possible for worms to secrete such a substance internally?' (Rhodes 1954, p. 449).

The phosphatic composition of conodont elements was recognized early in the history of their study. Pander (1856) believed initially that they were formed of calcium carbonate, but later detected calcium phosphate, which Harley (1861) showed to be the more abundant constituent. Ellison (1944) demonstrated that the crystal structure was of the apatite type, and subsequent analyses (Hass and Lindberg 1946, Phillips, In: Rhodes 1954, Rhodes and Wingard 1957, Pietzner *et al*. 1968) confirmed this, although producing differing conclusions about the members of the isomorphous apatite series involved. Pietzner *et al*. (1968) produced the most detailed formula,

and regarded the mineralogy of Devonian elements to be of the carbonate apatite, francolite. Rhodes (1954) considered the composition to be essentially similar to that of the bony armour of Devonian fish. This, together with the 'bone-like' nature of the basal body and the tooth-like appearance of the crown, provided a major line of evidence for those who favoured a fish affinity (e.g. Ulrich and Bassler 1926, Branson and Mehl 1933, Schmidt 1934, Ellison 1944). However, Gross (1954, 1957) pointed out that conodont elements grew by the accretion of successive lamellae on the outer surface of the unit, and are thus not homologous with the enamel teeth of vertebrates. Gross was still inclined towards a relationship of conodonts with fish, although he weakened the hypothesis still further by demonstrating that the basal body was not 'bone-like' as it lacked blood-vessel canals.

The idea that conodont elements shared affinities with worm jaws dates back to 1858 (Owen, In: Murchison 1859). The hypothesis gained further currency with the demonstration by Zittel and Rohon (1886) that the elements resembled the masticatory apparatuses of annelids, especially the echiuroids and the sipunculoids. Compositionally, though, annelid jaws are formed of chitin, and although some worms produce phosphatic tubes, none secretes calcium phosphate internally, which would be necessary to produce the external apposition of lamellae. A further argument is that annelid jaws are built by the addition of material to the inside of the base (Lindström 1964) and cannot grow in size like conodont elements. Lindström (1964, p. 120) concluded: 'The situation is evidently the same for the annelid hypothesis as for the fish hypothesis. It is false if applied to known, existing structures. Otherwise, it must be so vague as to be meaningless.'

The evident weaknesses in the hypotheses that conodont elements represented the teeth, jaws, or radulae of vertebrates, annelids, or molluscs led Lindström (1964) to develop the proposal previously argued by Hass (1941) that they functioned as internal skeletal supports. Hass forcefully pointed out three reasons why he believed the elements could not have functioned as ingestive aids:

(1) They were surrounded by tissue during growth, and could only have performed after they were completely grown and had erupted from the secreting medium.
(2) Numerous conodont elements of all growth stages show that parts of their structure were broken away and then restored during subsequent lamellar accretion.
(3) Elements commonly show no sign of wear, and there were no records of specimens worn in such a way that their condition could only be ascribed to use.

Rhodes (1954), however, reported the existence of elements showing apparent attrition on the oral surface, and pointed out that the hardness of apatite is sufficient to allow the elements to function as teeth without showing significant wear. He also noted that fossil polychaete jaws (scolecodonts) from the same strata as conodont elements rarely show signs of surface attrition, even though they are formed of a softer, chitinous material. With regard to the regeneration of broken elements, he emphasized that in this process the broken portion is almost always lost; if the element were embedded in tissue it is more likely that the two fractured portions would be re-fused. Lindström (1964) countered this by reasoning that broken parts could have been expelled from the body or resorbed. This, he felt, would be advantageous mechanically, since a healed fracture would be weaker than a rejuvenation that consisted entirely of a continuation of the outermost apatite lamellae.

The necessity for conodont elements to be covered by tissue throughout their growth persuaded Lindström (1964, p. 123) that there was no reason to assume that they were ever exposed during life. If, as Hass had suggested, they were skeletal supports, then what kind of organ did they strengthen? From examination of element morphologies, Lindström (1964, 1973, 1974) suggested that they might have supported tentacles within a food-gathering apparatus. Details of Lindström's hypothesis and its implications for conodont affinity will be discussed later.

Despite the difficulties highlighted by Hass (1941) and Lindström (1964), the idea that conodont elements functioned as teeth or grasping units has refused to die. Bengtson (1976) overcame the apparent paradox of internal growth but external function by postulating that euconodont elements were completely engulfed in epithelial pockets with the secreting tissue adherent only to the basal body. The crown could then be extruded from the pocket when in function and retracted during rest. The whole element could continue to grow between functional periods, and evidence of wear on the outer surface would not be expected. Furthermore, restoration of broken and lost cusp and denticle tips would be readily effected. These points were reinforced by Jeppsson (1979, 1980), who also demonstrated a number of close analogies between conodont elements and vertebrate teeth. Jeppsson (1979) was at pains to stress that these similarities did not imply any taxonomic relationship, but reflected the development by the two groups of similar structures to perform similar functions using the same material, apatite. He also noted that different types of bite were accomplished by teeth of different designs, and that gross differences in conodont elements could similarly be explained by their disparate functions. Thus teeth may occlude, which may be envisaged for many platform elements, or they may shear, for which ozarkodiniform elements have an appropriate shape, or they may grasp without meeting, which would suit many ramiform and coniform morphologies.

Jeppsson's 1979 paper led to a set of short discursive contributions in the same journal, *Lethaia*, as the original publication. Conway Morris (1980) professed a continued preference for the tentacle-support model, and argued that the peeling back of secretory tissue from even simple conodont elements would present a space problem, a point previously made by Nicoll (1977). This was countered by Bengtson (1980), who suggested that 'space problems' could equally be used to prove the impossibility of eversible proboscides or of the retractable claws of cats. If this debate proved one thing, it was that there was still no general agreement on the fundamental question as to whether conodont elements were exposed or tissue-covered during their function. In the view of Bengtson (1980), this 'baffling and sad fact' resulted in the entire field of functional morphology being essentially closed to conodont workers. Jeppsson (1979) had perhaps already shown that this was not strictly true, but there are inevitable limitations to interpretation, and even the discovery of preserved conodont animals has not resolved the situation. Hence, in this volume, papers supporting a tooth function (Aldridge *et al.* 1987, Smith *et al.* 1987) and a filtering function (Nicoll and Rexroad 1987, Nicoll 1987) continue to vie for acceptance.

## 1.3 THE ARCHITECTURE OF CONODONT APPARATUSES

Although much of the literature addressing the function of conodont elements has essentially treated them as separate entities, it is well-known that they served as parts of a multi-element skeletal apparatus that presumably functioned in an integrated manner. Pander's (1856) view was that all the teeth from one individual would be of similar morphology, and he considered it most unlikely that ribbed, keeled, smooth, and truncated elements could all have occurred in the same mouth. Hinde (1879), however, found a variety of forms preserved in association on a bituminous shale bedding surface of the Devonian Genesee Shale in New York. He considered that these all must have belonged to the same animal, and assigned the name *Polygnathus dubius* to the whole association. Hinde did not figure the assemblage, but used discrete specimens from various localities and horizons to illustrate the different forms incorporated in his concept of the species. Huddle (1972) re-studied Hinde's material and published photographs of two conodont-covered shale surfaces from the deposited collection. Huddle thought that these were two separate assemblages, but it seems more likely that they represent a single congregation of elements that had been separated onto two surfaces when the shale was split. In any event, the large number and chaotic orientation of the elements indicates that several individual animals are represented and the accumulation reflects some post-mortem process. Huddle (1972) expressed the opinion that a predator producing food balls might be the origin, and this seems as likely an explanation as any.

More compelling evidence of the nature of the conodont apparatus was published in 1934, when Schmidt and Scott independently described bedding plane assemblages from Carboniferous shales of Germany and Montana, respectively. A few years later, additional material from North American localities was reported by Scott (1942) and Du Bois (1943). Some of these associations of elements are rather disorganized, but patterns of consistent orientation were apparent in others (Fig. 1.2), and each of these evidently comprises the skeletal remains of a single conodont animal. More recently, many more Carboniferous assemblages have been recognized (Rhodes 1952, Schmidt and Müller 1964, Avcin 1974, Norby 1976, Rhodes and Austin 1985), and it is possible to recognize a limited variety of recurrent patterns of element arrangements. Assemblages that show more random or irregular arrangements may represent coprolitic or regurgitated aggregations, chance associations of elements, or post-mortem disturbances of single apparatuses.

The initial response to the discovery of natural assemblages was not entirely positive. Branson and Mehl (1936, 1938), for example,

Fig. 1.2—A bedding plane assemblage from the Carboniferous of Bailey Falls, La Salle County, Illinois, ×34. Specimen no. x-6377, University of Illinois. Originally illustrated by Du Bois (1943, pl. 25, fig. 21).

regarded most to be excretionary in nature, and observed that discrete elements in large collections did not occur in the relative proportions predicted by the assemblages. In their opinion, the weight of evidence was against a 'complex dental arrangement' (Branson and Mehl 1936, p. 233). However, assemblages displaying the same element types in the same numbers continued to be discovered, and, with Rhodes' review and classification of the known material in 1952, most workers became convinced of the multi-element nature of conodont apparatuses. Many of the Carboniferous assemblages are of taxa referable to the super-family Polygnathacea and display a consistent composition of six or seven different kinds of elements (Fig. 1.3). Most of these elements are paired, with two pairs of pectiniform (Pa and Pb) elements positioned close to a set of a ramiform (M, Sa, Sb, Sc, Sd) elements. In this ramiform 'basket' there are two dolabrate (M) elements, four bipennate (Sc) elements, two digyrate (Sb) elements, two sharply bent bipennate (Sd) elements, and a single, axial, alate (Sa) element. The bedding plane assem-

blage of the related genus *Ozarkodina*, described from the Lower Devonian of the Soviet Union by Mashkova (1972) shows a comparable arrangement, but lacks the Sd elements. Other conodont assemblages are known from the Devonian (Nicoll 1977, Puchkov *et al*. 1982, J. Zikmundova, pers. comm.) and from the Triassic (Rieber 1980), but the vast majority of specimens are from Carboniferous strata. Genera not referred to the Polygnathacea are represented by a few of the assemblages, but most interpretations of apparatus architecture and function have relied on the more common associations.

Natural assemblages of conodont elements are also found in the form of clusters, in which elements lying in juxtaposition in the sediment have become diagenetically fused together. These have then been freed from the rock during routine acid digestion for the recovery of conodont elements (Fig. 1.4). The nature of many clusters is more equivocal than that of the bedding plane assemblages, and several are probably of coprolitic origin. Others represent the partial or, very rarely, complete appar-

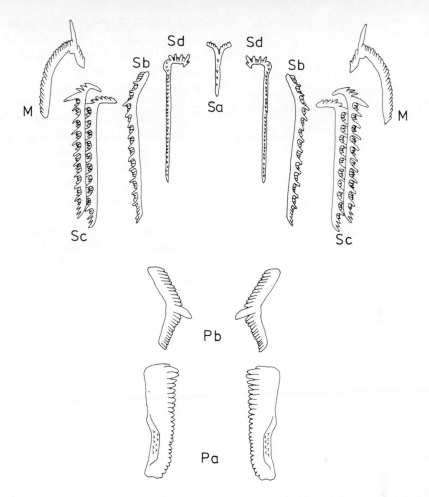

Fig. 1.3—Diagrammatic representation of the multielement composition of a Carboniferous polygnathacean assemblage, modified from Norby (1976, p. 103).

atuses of single individuals that have become
fused in their original alignments. Although
most clusters give information on the elemental compositions of apparatuses, they are generally less valuable than bedding plane assemblages in helping us to reconstruct the three-
dimensional structure of the conodont skeleton, and the evidence they provide must be
considered with care. The first clusters to be
described were of fused pectiniform pairs from
the Silurian of Indiana (Rexroad and Nicoll
1964), and they have subsequently been recognized in all conodont-bearing systems and all
over the world (e.g. Landing 1977, Repetski

1980, Nowlan 1979, Pollock 1969, Lange
1968, Nicoll 1982, Austin and Rhodes 1969,
Behnken 1975, Ramovs 1978).

It is important to note that conodont assemblages, both on bedding planes and as clusters,
are rare. Only a few hundred of each of the two
types have been found, and just a few localities
have produced the majority of specimens.
Most conodont taxa are known only from dis-
sociated elements, and the composition of their
apparatuses has to be deduced from distributional and morphological criteria (see e.g. Walliser 1964, Sweet and Bergström 1969, Jeppsson 1971). Hence, all natural assemblages are

of paramount importance for taxonomy, where they provide the only direct evidence of the multi-element nature of species, and for palaeobiology, where they provide the basis for assessments of the structure and operation of conodont apparatuses.

The assemblages figured by Scott (1934) show rather variable patterns of element orientation, and he did not use them at that time to develop a model of apparatus architecture. Schmidt (1934), in contrast, found polygnathacean assemblages showing a more regular alignment, and employed these in the construction of the first three-dimensional diagram of a multielement apparatus (Fig. 1.5A). The

Fig. 1.4—A fused cluster of ramiform conodont elements from the Carboniferous Granton Sandstones of Granton, Edinburgh, Scotland. Specimen no. IGSE 13823, Institute of Geological Sciences, Edinburgh, ×42.

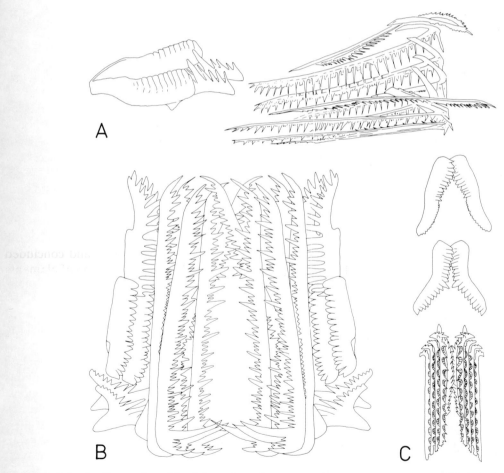

Fig. 1.5—Architectural models of the polygnathacean conodont apparatus: A, after Schmidt (1934); B, after Scott (1942); C, after Rhodes (1952).

essential points of his reconstruction are:

(a) All elements are longitudinally aligned, with the apparatus bilaterally symmetrical about a vertical, longitudinal plane.

(b) The arrangement is linear and elongate with the paired platform elements at one end and the ramiform group at the other; the intermediate angulate pectiniform elements are placed close to the platforms.

(c) The distal ends of the ramiform elements (conventionally posterior) point in the same direction as the cusps of the Pb elements and the platform ends of the Pa elements (also conventionally posterior).

(d) The dolabrate (M) elements are positioned on the flanks of the arched group of ramiform elements.

(e) The denticulated surfaces of the pectiniform elements face in the opposite direction to the denticulated surfaces of the ramiform elements.

In the absence of soft tissue associated with the assemblages, there was, of course, no direct evidence for the antero-posterior or dorso-ventral orientation of the skeleton. Schmidt (1934), however, regarded the elements to represent the mandibles, hyoid arch elements, and gill rakers of primitive placoderm fish, with the platform elements, therefore, at the anterior of the apparatus. A consequence of this view is that all elements were actually oriented in the reverse direction to that conventionally applied in descriptive work. It is also implicit from Schmidt's diagram that he considered the denticulated surfaces of the pectiniform elements to face dorsally.

The discovery of additional assemblages from Montana led Scott (1942) to propose a very different model. He felt that there was insufficient evidence to draw a wholly accurate picture, but produced the schematic arrangement shown in Fig. 1.5B. The apparatus is again interpreted as bilaterally symmetrical, but Scott placed the pectiniform elements directly alongside the ramiform group and oriented the platform elements in the opposing direction from the remainder. Scott's diagram also shows the denticulated surfaces of pectiniform and ramiform elements from the same side of the apparatus facing in the same (inward) direction. In his restudy of North American assemblages, Rhodes (1952) revised Scott's plan and produced a diagrammatic arrangement in which the elements form an elongate, linear series, as in Schmidt's model. However, Rhodes' arrangement (Fig. 1.5C) shows all the elements in the reverse relative orientation from that shown by Schmidt, and he additionally illustrated the dolabrate elements as the closest ramiforms to the axis of symmetry rather than the furthest from it. This interpretation of the apparatus pattern was probably strongly influenced by a single well-preserved assemblage collected from the Carboniferous of Illinois by Du Bois (1943, pl. 25, fig. 14; refigured by Rhodes 1952, pl. 126, fig. 11). This specimen is illustrated in Fig. 1.6, where the general pattern put forward by Rhodes is apparent, although the ramiform element group is somewhat disrupted. Rhodes did not directly attribute an anterior and a posterior to the apparatus, but the inference may be taken from his usage of conventional orientation in the descriptions of component elements that the platform elements are at the anterior. In a later contribution, Rhodes (1962) commented on the symmetry of conodont elements in assemblages, and concluded that it suggested lateral opposition of elements (as left and right forms) rather than an upper–lower jaw-like structure.

Although Rhodes (1952) stressed that the relative positions of components in his apparatus illustrations were purely diagrammatic, his putative arrangement has certainly influenced several authors' interpretations of how the conodont skeleton might have operated. This may partly be a result of the refiguring of his diagrams in widely consulted publications, including both editions of the *Treatise on Invertebrate Paleontology* (Rhodes 1962, Rhodes and Austin 1981). Prior to Rhodes' work, however, Du Bois (1943, p. 158), had

Fig. 1.6—Bedding-plane assemblage from the Carboniferous of Bailey Falls, La Salle County, Illinois, ×22. Specimen no. x-1480, University of Illinois. Originally illustrated by Du Bois (1943, p. 25, fig. 14).

and closing the anterior end. A primary sieving function was performed by the arched blade (Pb) elements, with secondary sieving carried out by the cusps of the ramiform elements. The elongate denticulated bars of the ramiform group were covered with ciliated tissue which could have served as a food trap and a respiratory organ. In this system, unlike Lindström's, only the ramiform elements were tissue-covered throughout life, while the functional surfaces of the Pa, Pb, and, perhaps, the M elements were exposed after development.

Although several authors supported the general idea of the conodont apparatus as a filtering system (e.g. Halstead 1969, Nicoll 1977), others continued to prefer a masticatory, grasping, or rasping function. Priddle (1974), for example, developed a suggestion by Schmidt (In: Schmidt and Müller 1964) that the elements could have borne horny cusps covering the entire unit, thus explaining their tooth-like form whilst being internally secreted. The apparatus, in Priddle's proposal, could have supported an array of such horny cusps set on an eversible lingual structure similar to that possessed by the extant myxinoids. Although the conventional arrangement of elements would not be necessitated by such a structure, it could readily be accommodated in the model. More direct analogies of apparatus structures were employed by Rietschel (1973), who compared the arrangements proposed by both Schmidt and Rhodes with the radulae of gastropods, the jaws of polychaetes, and the grasping apparatuses of chaetognaths. Rietschel's conclusion was that the chaetognaths provided the best comparative model for the conodonts. Dzik (1976), in contrast, recognised an analogy, and perhaps a homology, with the dermal denticles of thelodonts and fish.

There was, however, some doubt about the accuracy of Rhodes' reconstructed arrangement. Lange (1968), in describing a number of Upper Devonian clusters from the Kellwasserkalk, noted that where platform elements occurred in his material they were positioned to the posterior of the ramiform elements.

clearly held a similar picture in his mind when he argued that the elements formed a gradient 'in which the polygnathids perform the preliminary mastication, and the hindeodellids the final comminution or straining'. The same arrangement, but a totally different function, was adopted by Lindström (1964), who envisaged the whole apparatus as a supporting structure for tentacles, which filtered food particles from the water and passed them on to the intestinal canal. A filtering function was also favoured by Hitchings and Ramsay (1978), who based their interpretation directly on Rhodes' reconstructions. In their model, the conodont elements lined a feeding and respiratory channel with the platforms opening

Although he felt that these clusters were all probably coprolitic, they should retain some indication of the original relative positioning of elements, and the evidence they provided was more in agreement with Schmidt's reconstruction than with that of Rhodes. This conclusion was supported by Jeppsson (1971) who accepted that the ramiform elements were at the anterior and the platforms at the posterior. Jeppsson (1971, p. 109) also noted that evidence from assemblages and clusters indicated that the M elements were located on the outside of the apparatus, as proposed by Schmidt, rather than centrally. In his reconstruction of *Oulodus angulatus*, isolated from the gut of a palaeoniscid fish, Nicoll (1977) adopted a broadly similar arrangement, but with a pair of dolabrate elements set anterior to the elongate ramiform group.

It is apparent that, in the lack of any preserved soft tissue, studies of natural assemblages of conodont elements led to widely differing interpretations of apparatus architecture and function. The conflicting functional hypotheses that had developed from examination of isolated elements were not resolved by evidence of the complete skeletal structure. The only way in which the various proposals could be seriously tested was likely to be by the discovery of complete specimens of conodont animals, and a search for such fossils has been an important part of conodont palaeontology in recent years.

## 1.4 THE NATURE OF THE CONODONT ANIMAL

One obvious place to look for traces of soft tissue associated with conodont elements is in the beds that have produced the most numerous bedding-plane assemblages. Carbonaceous patches and impressions do occur in these shales, and Du Bois (1943, p. 156) noted an association of a hindeodellid element with a brown carbonaceous film that 'apparently represents a fossilized portion of the cuticle of some worm-like creature'. The structures that Du Bois interpreted as parapodia or cirri are, however, not clear from his illustration (Du Bois 1943, pl. 25, fig. 16), nor are similar impressions associated with complete, well-preserved assemblages. Those assemblages which are surrounded by a brownish film are commonly disorganized, and are most likely to be of coprolitic or similar origin. The asphaltic 'blebs' described by Scott (1969) from the Heath Shale of Montana are also almost certainly faecal in nature. Scott initially believed that they represented the cartilaginous portions of the heads of conodont animals, but the elements they contain are chaotically arranged, commonly broken, and do not constitute complete apparatuses. There is no evidence of soft tissue surrounding well-preserved, organized assemblages in any of the Carboniferous black shales.

It was in Carboniferous strata of Montana, though, that the first indisputable associations of soft-bodied organisms and conodont elements were discovered. During a routine check in 1969 of a fossil fish locality in the Bear Gulch Limestone, W. G. Melton Jr and J. J. Horner found carbonized impressions of animals that proved on laboratory examination to contain conodont elements (Melton 1971, 1972). These specimens were shown to a number of conodont specialists at the North American Paleontological Convention held that year in Chicago. Among those present was Dr Harold Scott, who had described the first North American bedding-plane assemblages in 1934. Scott was deeply impressed by the specimens, and announced to the Convention that the conodont animal had been found. Melton and Scott (1973) went on to collaborate on the publication of a full description of the new fossils, which included four complete animals.

The Bear Gulch animals are exquisitely preserved on bedding surfaces in fine-grained limestones. They are elongate, around 50–70 mm long by 12–15 mm high, with an anterior opening and a posterior fin (Fig. 1.7). In the original specimens, conodont elements

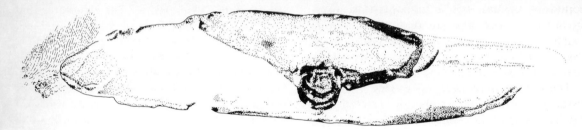

Fig. 1.7—A 'Conodontochordate' from the Bear Gulch Limestone of central Montana, ×2. Drawing of specimen no. UM 6027, University of Montana, Missoula; compiled from illustrations by Melton (1972) and Conway Morris (1985).

occur within the triangular gut, termed the 'deltaenteron' (Melton and Scott 1973), although in at least two specimens collected later (USNM 183566, see Scott 1973; UM 6099) they are absent. Melton and Scott (1973) believed that the conodont apparatus served to filter phytoplankton within this jawless animal, which they regarded as a free-swimming chordate. The relationships to other chordates were unclear, and Melton and Scott assigned the new animals to a new subphylum, the Conodontochordata.

These discoveries were undoubtedly of considerable palaeontological interest, but they were accorded a mixed reception by conodont workers. Lindström (1973, 1974) pointed out that the conodont elements within the animals are mostly disordered, and that only one specimen contains the remains of a nearly complete apparatus. He regarded the conodontochordates as conodont eaters, an opinion shared by, among others, Hitchings and Ramsay (1978) and Conway Morris (1976, 1985). Conway Morris (1985) has examined most of the known specimens and provided a particularly comprehensive critique, stressing the lack of conodont elements in some specimens and the presence of the remains of fish and other organisms in the guts of others (CM75–72306, UM 6101). He also noted the inordinately large number of elements in some specimens, for example UM 6030, in which he identified over eighty widely scattered elements including at least four Pa elements. In an undescribed

specimen numbered 72914 and currently held at the University of Wisconsin, Madison, the gut contains two patches of elements in which I have counted at least 16 Pa elements of various sizes. This is clearly at variance with our understanding of conodont apparatuses from bedding-plane assemblages, and the combined evidence overwhelmingly indicates that these animals were conodontophages.

Although the conodontochordates were not conodonts, they do provide us indirectly with some information about conodont palaeobiology and ecology. Lindström (1974) recognized this, and used the conodontophages to constrain the design of a hypothetical conodont animal, in which he envisaged the apparatus as the supporting structure for a tentaculate lophophore. The suggestion that the animal was a lophophorate had been floated earlier (Lindström 1964, 1973), but the fact that it was small enough to pass into the gut of a conodontochordate led Lindström to deduce that it was, at most, five or six times as long as the bars of the ramiform elements. In view of this, he considered (Lindström 1974, p. 740) that the animal was 'more likely to have been oblong or even barrel-shaped rather than long and worm-shaped'. Through consideration of bedding-plane assemblage patterns and the functional morphology of elements, Lindström further proposed that the elements encircled the mouth with their denticulated surfaces pointing outwards, thus serving as passive protection as well as lophophore supports. If the conodont animal was a lophophorate, then there are evident organizational similarities to inarticulate brachiopods, and Lindström (1973) suggested that there was a distinct possibility that the two groups shared a common ancestor.

There was, of course, no evidence of soft tissue to corroborate Lindström's model, but the hypothesis that conodonts were lophophorates was supported by Conway Morris (1976) on the basis of a unique specimen from the Middle Cambrian Burgess Shale of British Columbia. *Odontogriphus omalus* Conway Morris is a bilaterally symmetrical, compressed animal, with a looped structure towards one end (Fig. 1.8). The single fossil is represented by both part and counterpart, with different features of the loop apparent on each. On the counterpart, regularly-spaced depressed areas were interpreted by Conway Morris as retracted tentacles, while on the part are moulds of 20–25 small spikes which he believed were the remains of rigid supports for these tentacles. He interpreted the whole structure as a tentacular feeding apparatus that surrounded a mouth in life and was comparable with the lophophores of other lophophorate phyla. He also suggested, from two principal lines of reasoning, that *O. omalus* is an example of a conodont animal. Firstly, the spiky supports of the feeding apparatus bear a strong resemblance in outline and size to some simple Cambrian conodont elements, and, secondly, the whole lophophore shows considerable agreement with the hypothetical reconstruction proposed by Lindström (1974). The evidence, however, must be considered circumstantial. As only the moulds of the spiky elements are preserved, we cannot test if their composition and histology compare with those of the conodont group, and, while the convergence of ideas represented by Lindström's and Conway Morris's reconstructions is interesting, it certainly cannot be regarded as conclusive.

It now appears unlikely that *Odontogriphus* is closely related to the conodonts (Briggs *et al*. 1983, Briggs and Conway Morris 1986), and

the fossil, although interesting, has shed little real light on the nature of the Conodonta. More recently, though, discoveries have been made that have revealed the body plans of at least some conodont animals, and have provided the first firm foundations for discussions of their zoological affinities.

As part of an investigation of Lower Carboniferous shrimp beds in Scotland, Dr Euan Clarkson in early 1982 re-examined large collections of the Granton shrimp bed of Edinburgh that are housed at the British Geological Survey in Edinburgh. Among the material were two slabs, part and counterpart, that preserved an elongate worm-like organism with conodont elements in the head region. Dr Clarkson showed the specimen to Dr Derek Briggs and myself, and we were both convinced that this, at last, was truly a conodont animal. The specimen is 40.5 mm long and mostly less than 1.8 mm across, with a slightly expanded head region comprising two lobate structures flanking a central lumen (Fig. 1.9). The conodont apparatus lies behind this, with the ramiform group to the anterior and the Pa elements at the posterior; the paired pectiniform elements are oriented transverse to the long axis of the animal, while the internally aligned ramiform set are at a high oblique angle. This arrangement of elements is similar to that shown by many bedding-plane assemblages. The soft tissue of the trunk, preserved in vivianite, reveals a linear axial feature, repeated oblique structures that are probably segments, and, at the posterior end, ray-supported caudal and posterior fins on one side of the animal only. A full description of the specimen, assigned to *Clydagnathus*? cf. *cavusformis*, was given by Briggs *et al*. (1983), who recognized possible similarities to two extant groups, the chaetognaths and the chordates.

Chaetognaths are elongate, dorso-ventrally flattened worms with lateral fins (Fig. 1.10A). Their general features match well with those of the first Granton conodont animal if the latter is considered to be compacted in dorso-ventral aspect. The axial trace could then represent the

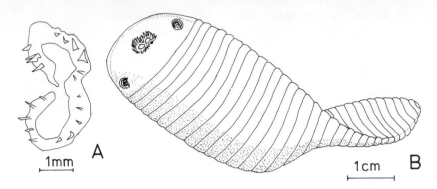

Fig. 1.8—*Odontogriphus omalus* Conway Morris: A, the feeding apparatus on the part, specimen no. USNM 196169, US National Museum of Natural History, Washington, D.C.; B, reconstruction of complete animal. Redrawn from Conway Morris (1976), with permission.

median mesentery that divides the trunk and tail coelom into lateral compartments, while chaetognaths also possess sets of chitinous grasping spines that flank the mouth and are covered at rest by a retractable hood. One problem with suggesting a chaetognath affinity, though, is that chaetognaths exhibit no structures that can be readily compared with the oblique segmentation evident along the trunk of the conodont.

Aspects of the Granton animal also invite comparison with simple chordates, such as amphioxus (*Branchiostoma*), lampreys, and myxinoids (Fig. 1.10B). The similarities are evident in this case if the trunk of the conodont is supposed to be compacted in lateral aspect, with the fins dorsal. The segmentation might thereby be homologous with the V-shaped myotomes in amphioxus, and the axial trace could represent a wall of the gut, or another linear feature such as the notochord. The preserved features of the soft tissue can thus be accommodated well in a chordate model.

The unique nature of conodont elements combined with uncertainties regarding the nature of the soft structures displayed by the Granton animal led Briggs *et al*. (1983) to conclude that, at that time, neither the chaetognath nor the chordate model provided a satisfactory solution to the problem of conodont affinity. We chose instead to follow the lead set in the *Treatise on Invertebrate Paleontology*

(Clark 1981) and to assign conodonts to a separate phylum, the Conodonta, until more specimens with preserved soft parts were discovered.

Other authors were not so wary. While most accepted the authenticity of the new conodont animal, very different views were expressed as to its nature. Bengtson (1983a) argued a close analogy between the feeding apparatuses of conodonts and chaetognaths, and considered that 'it may be more than a coincidence that the new conodont animal conforms in so many aspects to chaetognath anatomy'. Others (Janvier 1983, Rigby 1983) preferred to stress the resemblances to vertebrates, with Rigby proposing a close relationship to the gnathostomes. A particularly comprehensive case for regarding conodonts as chordates was presented by Dzik (1986).

It was clear that these differences in interpretation could only be fully resolved by additional evidence, and an assiduous search of the Granton shrimp bed was mounted. This was rewarded in 1984 by the discovery of three more specimens (Fig. 1.11). collected by Neil Clark and prepared and described by Aldridge *et al*. (1986). On one of these specimens only the posterior part of an animal is preserved, but the other two display associated conodont apparatuses in the anterior portion and served to dispel any possible doubts that the original specimen is genuinely a conodont.

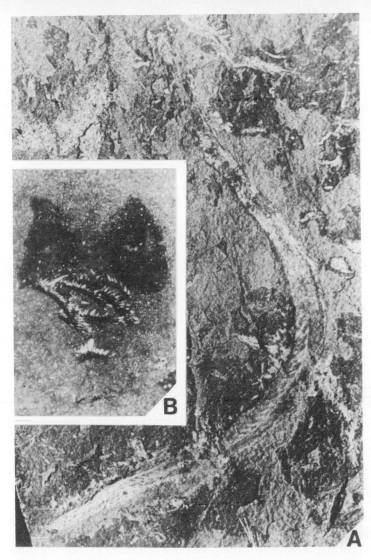

Fig. 1.9—*Clydagnathus*? cf. *cavusformis* Rhodes, Austin and Druce, from the Granton shrimp bed, Edinburgh: A, the complete specimen, ×7, IGSE 13821 (part), British Geological Survey, Edinburgh; B, head region of counterpart, ×19, IGSE 13822.

All of the new specimens provide some additional detail of soft-part structures, and in particular confirm that the trunk of the animal is segmented, with clear evidence of V-shaped somites. One specimen shows closely-spaced fin rays on both sides of the tail, extending further along the trunk on one side than the other, and all three show axial lines, rather more complex than in the first specimen but still equivocal in nature. The new data were regarded by Aldridge *et al*. (1986) as sufficient to negate the possibility of chaetognath affinity, and, although still in need of some clarification, appeared consistent with a chordate model. The most attractive hypothesis currently is that the conodonts represent a separate group of jawless craniates (Aldridge and Briggs 1986, Aldridge *et al*.

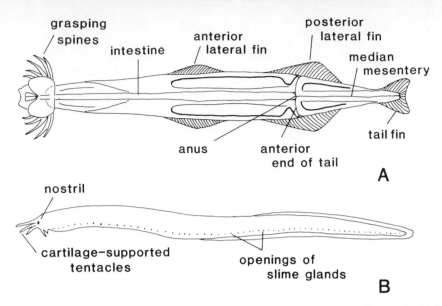

Fig. 1.10—A, diagram of a modern chaetognath, dorsal view, $\frac{1}{4}$; B, diagram of a modern myxinoid, lateral view, $\times\frac{1}{4}$.

1986), perhaps with some relationship to the myxinoids (Fig. 1.12).

No known chordates possess a skeletal structure that can with absolute confidence be homologized with the euconodont apparatus. The lingual structure of hagfish is broadly comparable (Priddle 1974, Dzik 1986, Aldridge *et al.* 1986), but the 'teeth' of *Myxine* comprise horny cusps mounted on cartilaginous supporting plates and the similarities may only be superficial. In the lack of any directly comparable structures, we cannot use our perception of conodont affinity to resolve the question of conodont element function. The two possibilities remain that they performed as teeth or that they supported a tentacular feeding/respiratory organ. Briggs *et al.* (1983) considered that an integrated system with the anterior ramiform elements grasping prey to be processed by the posterior pectiniform elements provided the more credible alternative. This view was endorsed by, among others, Bengtson (1983a), Aldridge and Jeppsson (1984), Sweet (1985), and Dzik (1986), and is further argued for in Chapter 4 of this volume (Aldridge *et al.* 1987).

However, not all conodont workers are convinced by this functional interpretation, and the alternative tissue-support idea is still strongly urged by Nicoll (1985), in particular. Nicoll (1985) employed evidence from beautifully preserved Upper Devonian clusters, together with the first Granton specimen, to propose a model in which the elements were linearly arranged along a ventral food groove in the head of the animal. He envisaged that the anterior group of ramiform elements, covered with ciliated tissue, would act as a sieve to sort particulate food matter which was then gently crushed by the posterior pectiniform elements prior to entry into the gut. Aldridge and Briggs (1986) commented that the Pa and Pb elements are rather large and robust in relation to the food particles that would be captured by such a sieve, and that their efficiency in comminuting food would be diminished by the postulated covering of soft tissue. However, in Chapter 3 (Nicoll and Rexroad 1987) and Chapter 5 (Nicoll 1987) of this volume it is still contended that the apparatus could have operated as a tissue-covered food-gathering mechanism.

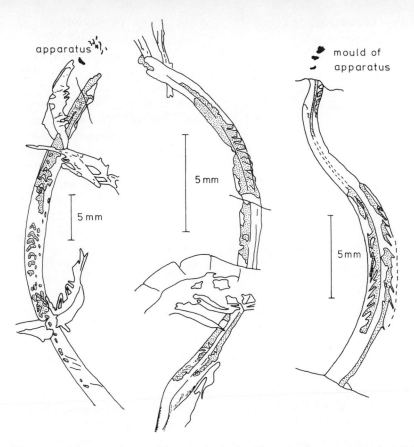

apparatus

mould of
apparatus

5 mm

5 mm

5 mm

Fig. 1.11—Diagrams of three specimens of the conodont animal from the Granton shrimp bed, Edinburgh.
The remains of shrimps lie across the left-hand and central specimens.

Much of the discussion of the function of conodont apparatuses has centred on polygnathacean taxa, which are by far the most abundantly represented group in natural assemblages. However, it is pertinent to consider how the different functional models might be applied to apparatuses of different structures which may comprise elements of very different morphological styles. In the Hibbardellacea, for example, all the elements, including those assigned to the Pa and Pb positions, are of ramiform morphology with those of each species showing internal consistency in process form and denticulation. It is unlikely that these elements were widely differentiated in function, and Aldridge and Briggs (1986) suggested that the entire apparatus may have served to grasp prey which was consumed without further processing. A similar interpretation could apply, of course, if the whole apparatus is presumed to have acted as a sieve. In *Icriodus*, and perhaps in other members of the Icriodontidae, the apparatus comprises a pair of scaphate Pa elements and a suite of cones, of which there may have been more than 140 in each individual (Nicoll 1982). Nicoll (1982) suggested that each of these cones might be homologous with a single denticle in other apparatuses, and that the *Icriodus* apparatus could thus be functionally very similar to those of polygnathacean genera.

The apparatuses most remote morphologically from those of the polygnathaceans are those comprised entirely of coniform elements.

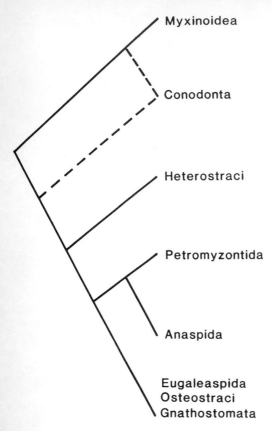

Myxinoidea

Conodonta

Heterostraci

Petromyzontida

Anaspida

Eugaleaspida
Osteostraci
Gnathostomata

Fig. 1.12—Cladogram illustrating two possible hypotheses of the relationships between the Conodonta and other craniates; after Aldridge *et al.* (1986), who provided a full discussion of the characters considered.

These showed considerable diversity in the early part of euconodont history and have been considered to have their closest functional analogues in the grasping apparatuses of chaetognaths (Aldridge *et al.* 1985, Dzik and Drygant 1985, 1986, Dzik 1986). Some coniform conodont elements have sharp edges, keels or costae, and it was suggested by Aldridge and Briggs (1986) that these may have performed a cutting or shearing function in addition to grasping. Understanding of the architecture and function of coniform apparatuses has been hampered by the relative lack of good natural assemblages, although several clusters have been described (e.g. Barnes 1967, Lange 1968, Pollock 1969, Now-

lan 1979, Aldridge 1982). More recently, Dzik and Drygant (1986) documented an almost complete assemblage of *Panderodus* from the Llandovery of Ukraine, and, of particular importance, Mikulic *et al.* (1985a,b) reported a conodont animal of the same genus from strata of similar age in Waukesha, Wisconsin. Although neither the soft tissue nor the skeletal apparatus of this animal is perfectly preserved, the specimen throws considerable light on the arrangement and function of coniform apparatuses. The Waukesha animal is further illustrated and described by Smith *et al.* (1987) in Chapter 6 of this volume, where a discussion of the architecture of panderodontid apparatuses is developed.

## 1.5   THE ORIGIN AND EVOLUTION OF CONODONTS

Euconodont elements, consisting of a centrifugally-secreted crown and a basal body, first appeared in the late Cambrian. Bengtson (1976, 1983b) proposed that their ancestry lay in the protoconodonts, from which he suggested they developed through an intermediate paraconodont state (Fig. 1.13). This sequence was presumed to involve gradual retraction of the elements into epithelial pockets, from which the crown of euconodonts could be exposed when the apparatus was in use. A striking morphological and structural similarity between protoconodont elements and the grasping spines of extant chaetognaths was documented by Szaniawski (1982, 1983), who regarded this as evidence of a relationship between the two, although not proof of identity. Others (e.g. Sweet 1985) felt it likely that protoconodont elements are the grasping spines of chaetognaths, and Bengtson (1983b, p. 5) noted that, if his postulated ancestry for euconodonts were true, then 'paracondont and eucondont animals may represent a branch of the chaetognaths that had developed pharyngeal denticulation.' Further histological support for this contention is provided by Szaniawski

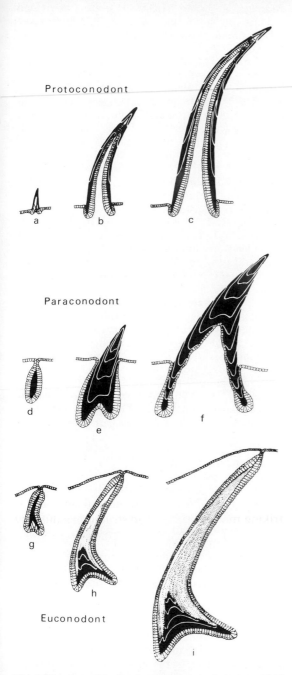

Protoconodont

a    b    c

Paraconodont

d    e    f

Euconodont

g    h    i

Fig. 1.13—A model of early conodont evolution, in which the differing internal structures of protoconodont, paraconodont and euconodont elements are explained by increased epithelial cover during growth (after Bengston 1976).

(1987) in Chapter 2 of this volume. However, an evolutionary link between proto-conodonts and paraconodonts remains to be conclusively established, although there is strong morphological, histological, and stratigraphical evidence that eucondonts evolved from paracondonts by acquiring a crown (Bengtson 1983b).

An alternative hypothesis for the origin of conodonts involves phosphatic cones of the genus *Fomitchella*, which has been found in Lower Cambrian strata in various parts of the world. Bengtson (1983b) illustrated specimens and demonstrated that the cones are constructed of lamellae that appear to have been added sequentially on the outer surface of the units. Thus, they compare with the mode of secretion of the euconodont crown, and it is conceivable that euconodonts may have evolved from *Fomitchella* or a close relative and not from paraconodonts. Bengtson (1983b, p. 12), however, preferred to consider the resemblances between *Fomitchella* and euconodonts as convergent or fortuitous, citing the lack of a known basal body and the random, finely granular apatite crystallities in the lamellae of the former as dissimilarities. In contrast, Dzik (1986) believed that it is plausible to derive the regularly oriented crystallites of euconodont crown lamellae from the pattern shown by *Fomitchella*. Thus, the question of euconodont origins remains open.

The development of the eucondont condition, from whatever ancestry, certainly gave rise to an impressive radiation of conodonts in the early Ordovician (Barnes *et al*. 1979, Miller 1984). A wide variety of skeletal apparatus plans were developed, some of which did not survive beyond the Ordovician. Others became stabilized in the later part of the period and gave rise to the successful stocks of younger times. Conodont evolution has been summarized recently by Sweet and Bergström (1981) and Sweet (1985), and aspects of evolutionary patterns shown by Ordovician to Triassic faunas are further discussed in Chapters 8–11 of this volume (Ethington *et al*. 1987, Jeppsson

1987, Ziegler and Lane 1987, Clark 1987).

Conodonts appear to have become extinct in the latest Triassic. The pattern of events leading up to their final demise has been reviewed by Clark (1983), who develops the theme in Chapter 11 (Clark 1987). The causes of the decline are speculative, but may be reflected in environmental changes caused by gradual transgression in the Late Triassic.

Of course, it is conceivable that after the last occurrence of conodont elements in the Triassic, the group continued to live on without producing mineralized, fossilizable hard parts. One recent report (Domingo 1983, p. 67) does, indeed, claim the discovery of a living conodont in a cave in Lanzarote. However, the leader of the expedition to the Jameos del Agua, Prof. Dr H. Wilkens (pers. comm. 1983), does not confirm this record, and notes that only primitive crustaceans were found. Thus, there is still no evidence for the existence of post-Triassic conodonts.

## ACKNOWLEDGEMENTS

My work on conodont palaeobiology is funded by N.E.R.C. Research Grant GR/3/5105. Unpublished data were kindly made available by Dr J. Dzik, Zakład Paleobiologii, Polish Academy of Sciences, Warsaw and Dr J. Zikmundova, Geological Survey, Prague; Dr R. L. Austin, University of Southampton, brought the report of a 'living conodont' to my attention. The paper was improved by comments from Dr D. E. G. Briggs and Dr M. P. Smith. Line drawings were drafted by Mrs J. Wilkinson, and photographic assistance was provided by Dr M. P. Smith and Mr A. Swift.

## REFERENCES

Aldridge, R. J. 1982. A fused cluster of coniform conodont elements from the late Ordovician of Washington Land, Western North Greenland. *Palaeontology* **25** 425–430, pl. 44.

Aldridge, R. J. and Briggs, D. E. G. 1986. Conodonts. In: A. Hoffman and M. H. Nitecki (eds): *Problematic fossil taxa*, Oxford University Press.

Aldridge, R. J., Briggs, D. E. G., Clarkson, E. N. K., and Smith, M. P. 1986. The affinities of conodonts – new evidence from the Carboniferous of Edinburgh, Scotland. *Lethaia* **19** 279–291.

Aldridge, R. J., Briggs, D. E. G., and Smith, M. P. 1985. The structure and function of panderodontid conodont apparatuses. In: R.J. Aldridge, R. L. Austin, and M. P. Smith (eds): *Fourth European Conodont Symposium (ECOS IV), Nottingham 1985, Abstracts*, University of Southampton, 2.

Aldridge, R. J. and Jeppsson, L. 1984. Ecological specialists among Silurian conodonts. *Special Papers in Palaeontology* **32** 141–149.

Aldridge, R. J., Smith, M. P., Norby, R. D., and Briggs, D. E. G. 1987. The architecture and function of Carboniferous polygnathacean conodont apparatuses. In: R. J. Aldridge (ed.): *Palaeobiology of Conodonts*, Ellis Horwood, Chichester, Sussex 63–75.

Austin, R. L. and Rhodes, F. H. T. 1969. A conodont assemblage from the Carboniferous of the Avon Gorge, Bristol. *Palaeontology* **12** 400–405.

Avcin, M. J. 1974. Des Moinesian conodont assemblages from the Illinois Basin. *Unpublished PhD thesis*, University of Illinois at Urbana-Champaign, 152 pp. 5 pls.

Barnes, C. R. 1967. A questionable conodont assemblage from Middle Ordovician limestone, Ottawa, Canada. *Journal of Paleontology* **41** 1557–1560.

Barnes, C. R., Kennedy, D. J., McCracken, A. D., Nowlan, G. S., and Tarrant, G. A. 1979. The structure and evolution of Ordovician conodont apparatuses. *Lethaia* **12** 125–151.

Behnken, F. H. 1975. Leonardian and Guadalupian (Permian) conodont biostratigraphy in Western and Southwestern United States. *Journal of Paleontology* **49** 284–315 2 pls.

Bengtson, S. 1976. The structure of some Middle Cambrian conodonts, and the early evolution of conodont structure and function. *Lethaia* **9** 185–206.

Bengtson, S. 1980. Conodonts: the need for a functional model. *Lethaia* **13** 320.

Bengtson, S. 1983a. A functional model for the conodont apparatus. *Lethaia* **16** 38.

Bengtson, S. 1983b. the early history of the Conodonta. *Fossils and Strata* **15** 5–19.

Branson, E. B. and Mehl, M. G. 1933. Conodont studies number one – Introduction. *The University of Missouri Studies* **8** 5–18.

Branson, E. B. and Mehl, M. G. 1936. Geological affinities and taxonomy of conodonts (abstract).

*Geological Society of America Proceedings for 1935*, 436; also *Pan-American Geologist* **65** 233.

Branson, E. B. and Mehl, M. G. 1938. Conodont assemblages (abstract). *Geological Society of America Proceedings for 1937*, 270.

Briggs, D. E. G., Clarkson, E. N. K., and Aldridge, R. J. 1983. The conodont animal. *Lethaia* **16** 1–14.

Briggs, D. E. G. and Conway Morris, S. 1986. Problematica from the Middle Cambrian Burgess Shale. In: A. Hoffman and M. H. Nitecki (eds): *Problematic fossil taxa*, Oxford University Press.

Clark, D. L. 1981. Classification. In: R. A. Robison (ed.): *Treatise on Invertebrate Paleontology, Part W, Supplement 2, Conodonta*, Geological Society of America and University of Kansas Press, Lawrence, Kansas, W102–W103.

Clark, D. L. 1983. Extinction of conodonts. *Journal of Paleontology* **57** 652–661.

Clark, D. L. 1987. Conodonts: the final fifty million years. In: R. J. Aldridge (ed.): *Palaeobiology of Conodonts*, Ellis Horwood, Chichester, Sussex, 165–174.

Conway Morris, S. 1976. A new Cambrian lophophorate from the Burgess Shale of British Columbia. *Palaeontology* **19** 199–222, pls. 30–34.

Conway Morris, S. 1980. Conodont function: fallacies of the tooth model. *Lethaia* **13** 107–108.

Conway Morris, S. 1985. Conodontophorids or conodontophages? A review of the evidence on the 'Conodontochordates' from the Bear Gulch Limestone (Namurian) of Montana, U.S.A. In: J. T. Dutro Jr and H. W. Pfefferkorn (eds.): *Palaeontology, Paleoecology, Paleogeography, Compte Rendu, Vol. 5, Neuvième Congrès international de stratigraphie et de géologie du Carbonifère, Washington and Champaign-Urbana, 1979*, Southern Illinois University Press, 473–480, 2 pls.

Denham, R. L. 1944. Conodonts. *Journal of Paleontology* **18** 216–218.

Domingo, X. 1983. Fosiles vivos en Lanzarote. *Cambio* **16** no. 600, 62–67.

Du Bois, E. P. 1943. Evidence on the nature of conodonts. *Journal of Paleontology* **17** 155–159, pl. 25.

Dzik, J. 1976. Remarks on the evolution of Ordovician conodonts. *Acta Palaeontologica Polonica* **21** 395–455, pls. 41–44.

Dzik, J. 1986. Chordate affinities of the conodonts. In: A. Hoffman and M. H. Nitecki (eds.): *Problematic fossil taxa*, Oxford University Press.

Dzik, J. and Drygant, D. 1985. The apparatus of panderodontid conodonts. In: R. J. Aldridge, R. L. Austin and M. P. Smith (eds.): *Fourth European Conodont Symposium (ECOS IV), Nottingham 1985, Abstracts*, University of Southampton, 9.

Dzik, J. and Drygant, D. 1986. The apparatus of panderodontid conodonta. *Lethaia* **19** 133–141.

Ellison, S. P., Jr. 1944. The composition of conodonts. *Journal of Paleontology* **18** 133–140.

Ethington, R. L., Engel, K. M., and Elliott, K. L. 1987. An abrupt change in conodont faunas in the Lower Ordovician of the Midcontinent Province. In: R. J. Aldridge (ed.): *Palaeobiology of Conodonts*, Ellis Horwood, Chichester, Sussex, 111–127.

Fahlbusch, K. 1964. Die Stellung der Conodontida im biologischen System. *Palaeontographica*, Abt. A **123** 137–201, pls. 16–22.

Globensky, Y. 1970. The nature of conodonts. *Le Naturaliste Canadien* **97** 213–228.

Gross, W. 1954. Zur Conodonten-Frage. *Senckenbergiana lethaea* **35** 73–85, 5 pls.

Gross, W. 1957. Über die Basis der Conodonten. *Paläontologische Zeitschrift* **31** 78–91, pls. 7–9.

Halstead, L. B. 1969. *The pattern of vertebrate evolution*, Oliver and Boyd, Edinburgh, 209 pp.

Harley, J. 1861. On the Ludlow Bone-bed and its Crustacean remains. *Quarterly Journal of the Geological Society, London* **17** 542–552, pl. 62.

Hass, W. H. 1941. Morphology of conodonts. *Journal of Paleontology* **15** 71–81, pls. 12–16.

Hass, W. H. and Lindberg, M. L. 1946. Orientation of crystal units of conodonts. *Journal of Paleontology* **20** 501–504.

Hinde, G. J. 1879. On conodonts from the Chazy and Cincinnati group of the Cambro-Silurian, and from the Hamilton and Genesee-shale divisions of the Devonian in Canada and the United States. *Quarterly Journal of the Geological Society, London* **35** 351–369, pls. 15–17.

Hitchings, V. H. and Ramsay, A. T. S. 1978. Conodont assemblages: a new functional model. *Palaeogeography, Palaeoclimatology, Palaeoecology* **24** 137–149.

Huddle, J. W. 1934. Conodonts from the New Albany shale of Indiana. *Bulletins of American Paleontology* **21** no. 72, 136 pp., 12 pls.

Huddle, J. W. 1972. Historical introduction to the problem of conodont taxonomy. *Geologica et Palaeonotologica* **SB1** 3–16, 1 pl.

Janvier, P. 1983. L''animal-conodonte' enfin demasqué? *La Recherche* **14** no. 145, 832–833.

Jeppsson, L. 1971. Element arrangement in conodont apparatuses of *Hindeodella* type and in similar forms. *Lethaia* **4** 101–123.

Jeppsson, L. 1979. Conodont element function. *Lethaia* **12** 153–171.

Jeppsson, L. 1980. Function of the conodont elements. *Lethaia* **13** 228.

Jeppsson, L. 1987. Lithological and conodont distributional evidence for episodes of anomalous oceanic conditions during the Silurian. In: R. J. Aldridge (ed.): *Palaeobiology of Conodonts*, Ellis Horwood, Chichester, Sussex, 129–145.

Landing, E. 1977. 'Prooneotodus tenuis' (Müller 1959) apparatuses from the Taconic allochthon, eastern New York; construction, taphonomy and the protoconodont 'supertooth' model. *Journal of Paleontology* **51** 1072–1084, 2 pls.

Lange, F. G. 1968. Conodonten-Gruppenfunde aus Kalken des tieferen Oberdevon. *Geologica et Palaeontologica* **2** 37–57, 6 pls.

Lindström, M. 1964. *Conodonts*, Elsevier, Amsterdam, 196 pp.

Lindström, M. 1973. On the affinities of conodonts. In: F. H. T. Rhodes (ed.): *Conodont paleozoology*. *Geological Society of America Special Paper* **141** 85–102.

Lindström, M. 1974. The conodont apparatus as a food-gathering mechanism. *Palaeontology* **17** 729–744.

Mashkova, T. V. 1972. *Ozarkodina steinhornensis* (Ziegler) apparatus, its conodonts and biozone. *Geologica et Palaeontologica* **S51** 81–90, 2 pls.

Melton, E. G. Jr 1971. The Bear Gulch fauna from central Montana. *Proceedings of the North American Paleontological Convention, Chicago, September 1969* **1** 1202–1207.

Melton, W. G. Jr 1972. The Bear Gulch Limestone and the first conodont bearing animals. *Montana Geological Society, 21st annual field conference*, 65–68.

Melton, W. G. Jr and Scott, H. W. 1973. Conodont-bearing animals from the Bear Gulch Limestone, Montana. *Geological Society of America Special Paper* **141** 31–65.

Mikulic, D. G., Briggs, D. E. G., and Kluessendorf, J. 1985a. A Silurian soft-bodied biota. *Science* **228** 715–717.

Mikulic, D. G., Briggs, D. E. G., and Kluessendorf, J. 1985b. A new exceptionally preserved biota from the Lower Silurian of Wisconsin, U.S.A. *Philosophical Transactions, The Royal Society, London*, Series B **311** 75–85.

Miller, J. F. 1984. Cambrian and earliest Ordovician conodont evolution, biofaces, and provincialism. In: D. L. Clark (ed.): *Conodont biofacies and provincialism. Geological Society of America Special Paper* **196** 43–68.

Müller, K. J. 1981. Zoological affinities of conodonts. In: R. A. Robison (ed.): *Treatise on Invertebrate Paleontology, Part W, Supplement 2, Conodonta*, Geological Society of America and University of Kansas Press, Lawrence, Kansas, W78–W82.

Müller, K. J. and Nogami, Y. 1971. Über den Feinbau der Conodonten. *Memoirs of the Faculty of Science, Kyoto University, Geology and Mineralogy Series* **38** 1–88, pls. 1–22.

Murchison, R. I. 1854. *Siluria. The history of the oldest known rocks containing organic remains with a brief description of the distribution of gold over the earth.* xvi + 523 pp., 37 pls. (3rd edition, 1859; 4th edition, 1867; 5th edition, 1872).

Nicoll, R. S. 1977. Conodont apparatuses in an Upper Devonian palaeoniscid fish from the Canning Basin, Western Australia. *BMR Journal of Australian Geology and Geophysics* **2** 217–228.

Nicoll, R. S. 1982. Multielement composition of the conodont *Icriodus expansus* Branson & Mehl from the Upper Devonian of the Canning Basin, Western Australia. *BMR Journal of Australian Geology and Geophysics* **7** 197–213.

Nicoll, R. S. 1985. Multielement composition of the conodont species *Polygnathus xylus xylus* Stauffer, 1940 and *Ozarkodina brevis* (Bischoff & Ziegler, 1957) from the Upper Devonian of the Canning Basin, Western Australia. *BMR Journal of Australian Geology and Geophysics* **9** 133–147.

Nicoll, R. S. 1987. Form and function of the Pa element in the conodont animal. In: R. J. Aldridge (ed.): *Palaeobiology of Conodonts*, Ellis Horwood, Chichester, Sussex, 77–90.

Nicoll, R. S. and Rexroad, C. B. 1987. Re-examination of Silurian conodont clusters from northern Indiana. In: R. J. Aldridge (ed.): *Palaeobiology of Conodonts*, Ellis Horwood Chichester, Sussex, 49–61.

Norby, R. D. 1976. Conodont apparatuses from Chesterian (Mississippian) strata of Montana and Illinois. *Unpublished PhD thesis*, University of Illinois at Urbana-Champaign, 303 pp., 21 pls.

Nowlan, G. S. 1979. Fused clusters of the conodont genus *Belodina* Ethington from the Thumb Mountain Formation (Ordovician), Ellesmere Island, District of Franklin. *Current Research, Part A, Geological Survey of Canada*, Paper 79–1A, 213–218.

Pander, C. H. 1856. *Monographie der fossilen Fische des Silurischen Systems der Russisch – Baltischen Gouvernements*, Akademie der Wissenschaften, St Petersburg, 1–91, 7 pls.

Pietzner, H., Vahl, J., Werner, H., and Ziegler, W. 1968. Zur chemischen Zusammensetzung und Mikromorphologie der Conodonten, *Palaeontographica* **128** 115–152, pls. 18–27.

Pollock, C. A. 1969. Fused Silurian conodont clusters from Indiana. *Journal of Paleontology* **43** 929–935, pls. 110–112.

Priddle, J. 1974. The function of conodonts. *Geological Magazine* **111** 255–257.

Puchkov, V. N., Klapper, G., and Mashkova, T. V. 1982 (dated 1981). Natural assemblages of *Palmatolepis* from the Upper Devonian of the northern Urals. *Acta Palaeontologica Polonica* **26** 281–298, pl. 25.

Ramovs, A. 1978. Mitteltriassiche Conodontenclusters in Slowenien, NW Jugoslawien. *Paläontologische Zeitschrift* **52** 129–137.

Repetski, J. E. 1980. Early Ordovician fused conodont clusters from the western United States. In: H. P. Schönlaub (ed.): *Second European Conodont Symposium (ECOS II), Guidebook, Abstracts, Abhandlungen der Geologischen Bundesanstalt,* **35** 207–209.

Rexroad, C. B. and Nicoll, R. S. 1964. A Silurian conodont with tetanus? *Journal of Paleontology* **38** 771–773.

Rhodes, F. H. T. 1952. A classification of Pennsylvanian conodont assemblages. *Journal of Paleontology* **26** 886–901, pls. 126–129.

Rhodes, F. H. T. 1954. The zoological affinities of the conodonts. *Biological Reviews, Cambridge Philosophical Society* **29** 419–452.

Rhodes, F. H. T. 1962. Recognition, interpretation, and taxonomic position of conodont assemblages. In: R. C. Moore (ed.): *Treatise on Invertebrate Paleontology, Part W, Miscellanea,* Geological Society of America and University of Kansas Press, Lawrence, Kansas, W70–W83.

Rhodes, F. H. T. and Austin, R. L. 1981. Natural assemblages of elements: interpretation and taxonomy. In: R. A. Robison (ed.): *Treatise on Invertebrate Paleontology, Part W, Supplement 2, Conodonta,* Geological Society of America and University of Kansas Press, Lawrence, Kansas, W68–W78.

Rhodes, F. H. T. and Austin, R. L. 1985. Conodont assemblages from the Carboniferous of Britain. In: J. T. Dutro Jr and H. W. Pfefferkorn (eds.): *Paleontology, Paleoecology, Paleogeography, Compte Rendu, Vol. 5, Neuvième Congrès International de stratigraphie et de géologie du Carbonifère, Washington and Champaign – Urbana, 1979,* Southern Illinois University Press, 287–300, 2 pls.

Rhodes, F. H. T. and Wingard, P. 1957. Chemical composition, microstructure and affinities of the Neurodontiformes. *Journal of Paleontology* **31** 448–454.

Rieber, H. 1980. Ein Conodonten-cluster aus der Grenzbitumenzone (Mittlere Trias) des Monte San Giorgio (Kt. Tessin/Schweiz). *Annalen des Naturhistorisches Museums Wien* **83** 265–274, 2 pls.

Rietschel, S. 1973. Zur Deutung der Conodonten. *Natur und Museum* **103** 409–418.

Rigby, J. K. Jr 1983. Conodonts and the early evolution of the vertebrates. *Geological Society of America Abstracts with Programs* **15**(6) 671.

Schmidt, H. 1934. Conodonten-Funde in ursprünglichen Zusammenhang. *Paläontologische Zeitschrift* **16** 76–85, pl. 6.

Schmidt, H. and Müller, K. J. 1964. Weitere Funde von Conodonten-Gruppen aus dem oberen Karbon des Sauerlandes. *Paläontologische Zeitschrift* **38** 105–135.

Scott, H. W. 1934. The zoological relationships of the conodonts. *Journal of Paleontology* **8** 448–455, pls. 58–59.

Scott, H. W. 1942. Conodont assemblages from the Heath Formation, Montana. *Journal of Paleontology* **16** 293–300, pls. 37–40.

Scott, H. W. 1969. Discoveries bearing on the nature of the conodont animal. *Micropaleontology* **15** 420–426, 1 pl.

Scott, H. W. 1973. New Conodontochordata from the Bear Gulch Limestone (Namurian, Montana). *Michigan State University, Paleontological Series* **1** 81–100, 3 pls.

Smith, M. P., Briggs, D. E. G., and Aldridge, R. J. 1987. A conodont animal from the lower Silurian of Wisconsin, U.S.A., and the apparatus architecture of panderodontid conodonts. In: R. J. Aldridge (ed.): *Palaeobiology of Conodonts*, Ellis Horwood, Chichester, Sussex, 91–104.

Sweet, W. C. 1985. Conodonts: those fascinating little whatzits. *Journal of Paleontology* **59** 485–494.

Sweet, W. C. and Bergström, S. M. 1969. The generic concept in conodont taxonomy. *Proceedings of the North American Paleontological Convention, Chicago, September 1969,* Part C, 157–173.

Sweet, W. C. and Bergström, S. M. 1981. Biostratigraphy and evolution. In: R. A. Robison (ed.): *Treatise on Invertebrate Paleonotology, Part W, Supplement 2, Conodonta,* Geological Society of America and University of Kansas Press, Lawrence, Kansas, W92–W101.

Szaniawski, H. 1982. Chaetognath grasping spines recognised among Cambrian protoconodonts. *Journal of Paleontology* **56** 806–810.

Szaniawski, H. 1983. Structure of protoconodont elements. *Fossils and Strata* **15** 21–27.

Szaniawski, H. 1987. Preliminary structural comparisons of protoconodont, paraconodont and euconodont elements. In: R. J. Aldridge (ed.): *Palaebiology of Conodonts*, Ellis Horwood, Chichester, Sussex, 35–47.

Ulrich, E. O. and Bassler, R. S. 1926. A classification of the toothlike fossils, conodonts, with descriptions of American Devonian and Mississippian species. *Proceedings of the United States National Museum* **68** Art. 12, 1–63, pls. 1–11.

Walliser, O. H. 1964. Conodonten des Silurs. *Abhandunglen des Hessischen Landesamtes für Bodenforschung* **41** 1–106, pls. 1–32.

Ziegler, W. and Lane, H. R. 1987. Cycles in conodont evolution from Devonian to mid-Carboniferous. In: R. J. Aldridge (ed.): *Palaeobiology of Conodonts*, Ellis Horwood, Chichester, Sussex, 147–163.

Zittel, K. A. and Rohon, J. V. 1886. Über Conodonten. *Sitzungsberichte der K. bayerischen Akademie der Wissenschaften zu Munchen* **16** 108–136, 2 pls.

# 2

# Preliminary structural comparisons of protoconodont, paraconodont, and euconodont elements

H. Szaniawski

## ABSTRACT

The microstructure and composition of paraconodont elements and of the basal bodies of primitive euconodont elements have been investigated using scanning and transmission electron microscopy. The close similarity to the histology of protoconodont elements supports suggestions of homology. A thin, translucent, coating occurs on the crown of primitive euconodont elements, and organic admixtures in the crown are illustrated for the first time. Two pairs of joined juvenile paraconodont elements are shown to compare with the teeth of modern chaetognaths, and a common ancestry for conodonts and chaetognaths is considered likely.

## 2.1 INTRODUCTION

Protoconodonts and paraconodonts have commonly been considered to be close relatives of the true conodonts (Müller 1959, 1962) and assignable to the same phylum. A hypothetical model of the evolution of euconodonts from protoconodonts was presented by Bengtson (1976), and I have recently described histological comparisons between Cambrian protoconodont elements and the grasping spines of recent chaetognaths (Szaniawski 1980, 1982, 1983). The similarity in shape between these two structures had been noted earlier by Müller and Andres (1976) who considered it to be a case of convergence, but the strong similarities in individual elements and the almost identical construction of apparatuses led me to interpret protoconodonts as close relatives of the Chaetognatha. Thus, the relationship between protoconodonts and euconodonts is of considerable biological importance and merits detailed investigation. The imprints of complete Carboniferous conodont animals discovered recently (Briggs *et al*. 1983, Aldridge *et al*. 1986) differ somewhat from recent chaetognaths, but do no exclude the possibility of a common origin. Unfortunately, to date there is only one described fossil specimen of a probable chaetognath animal (Schram 1973), and its food-capturing apparatus is not preserved. However, some indication of the relationships between conodonts and chaetognaths may possibly be derived from comparative histological studies of protoconodont, paraconodont, and primitive euconodont elements. The preliminary results of my investigations into this are presented in this chapter.

The material I have studied is from the Upper Cambrian of Poland, the Cambrian and

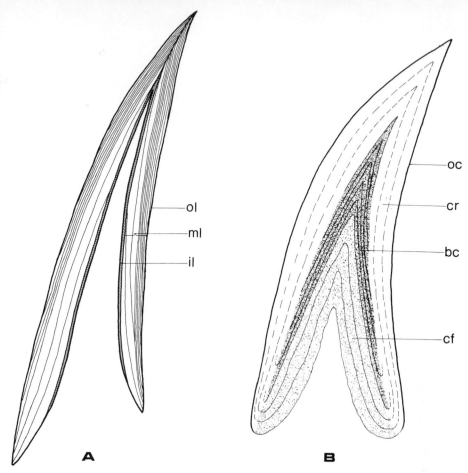

Fig. 2.1—Schematic longitudinal sections of a protoconodont element (A) and a primitive euconodont element (B). il – inner layer, ml – middle layer, ol – outer layer, bc – basal cone, cf – cone filling, cr – crown, oc – outer coating.

Tremadoc of Sweden, and the Tremadoc of Estonia. The investigation has involved examination of fresh and etched fracture surfaces, polished and etched sections of specimens embedded in plastic, thin sections, microtome slides, and ultramicrotome slides. The prepared specimens have been studied by optical microscopy, transmission electron microscopy (TEM), and scanning electron microscopy (SEM). Chromium sulphide and 3% nitric acid have been used for etching, and, in some cases, the critical point method has been used for drying etched specimens to avoid deformation. The material is preserved in the Institute of

Palaeobiology of the Polish Academy of Sciences in Warsaw, where most of the investigations were undertaken. Some of the studies were made in the Institute of Anatomy, University of Bergen, Norway, and others in the Institute of Palaeontology of Uppsala University, Sweden.

## 2.2 DESCRIPTIONS OF ELEMENT STRUCTURES

The structure of protoconodont elements has been described by Bengtson (1976) and Szaniawski (1982, 1983). They are built of

Fig. 2.2—*Furnishina* sp., Upper Cambrian, Uddagården, southern Sweden, ZPAL C.IV/35.1, ×135. Note translucent distal portion of cusp.

three layers (Fig. 2.1A), the outer of which is very thin, purely organic, and homogeneous. The middle layer is thick, laminated, and composed of organic matter and very fine apatite crystallites. This layer can be divided into two zones, with the inner zone containing less organic matter and being less regularly laminated than the outer one; the inner zone is also often secondarily mineralized. New lamellae were added from the inner side and grew towards the base. The third, innermost, layer is again thin, laminated, and mainly organic. It is constructed of a few very faint laminae which can only be recognized in very well preserved specimens.

The structure of protoconodont elements is very similar to that of Recent chaetognath grasping spines, except that the latter are of a purely organic, chitinous, composition. In addition to the comparisons documented previously (Szaniawski 1980, 1982, 1983), can be added the fact that the inner layer of grasping spines is also laminated (Bone *et al.* 1983, fig. 3G). I have independently observed such structures in spines of modern *Sagitta*, and the lamination is somewhat similar to that of the inner layer of protoconodont elements.

The structure of paraconodont elements was documented by Müller and Nogami (1971), but it is still possible to add some details. Etched longitudinal sections of specimens of *Furnishina* (Fig. 2.2; Pl. 2.1, fig. 5) show that they are constructed of more than 30 quite long lamellae. The lamellae are thickest centrally and show the direction of most rapid growth, whereas towards the margins the lamellae are narrow and dense. Some of the initially-secreted lamellae are probably circumferential. The basal cavity was formed during ontogeny by the invagination of some lamellae; juvenile specimens have no cavity at all (see also Szaniawski 1971, p. 404). Succeeding lamellae grew mainly towards the base and are discontinuous within the cavity, but display growth both at the outside of the element and inside the cavity. The earliest of these lamellae are extensive inside the cavity, but later ones were added mainly on the outside.

Transmission electron microscopy shows clearly that the paraconodont elements are composed predominantly of organic matter, with minor admixtures of phosphate. In ultrathin sections of demineralized specimens of *Furnishina* the organic matter is in very fine concentrations that are densely and apparently irregularly distributed within each of the lamellae (Pl. 2.1, figs 7, 8). However, in some sections the concentrations are linear, suggesting that the original structure of the organic matter was fibrous. Sections of specimens that have not been demineralized show that the phosphate occurs as extremely fine, nearly equidimensional crystallites (Pl. 2.1, fig. 9), apparently irregularly arranged in the lamellae. The similarity of the distribution of the crystallites in the non-demineralized sections to the distribution of the organic concentrations in the demineralized sections suggests that the crystallites incorporate most of the organic matter in the element.

Under the TEM it is possible to differentiate the lamellae within a single specimen, with the organic concentrations being more dense in some than others (Pl. 2.1, fig. 7). The lamellae secreted in the later period of element growth are often separated by thin, regular interlamellar spaces. At high magnifications, polished and etched surfaces of sections show a spongy structure (Pl. 2.1, fig. 6).

Paraconodont specimens from the Upper Cambrian of the Baltic region are usually black in colour and do not dissolve in strong hydrochloric acid. However, the cusps of some specimens from younger strata (above the *Agnostus pisiformis* Zone) are translucent and more phosphatic distally (Fig. 2.2), and dissolve more readily.

The structural similarity between paraconodont elements and the basal body of euconodont elements has been recognized for some time (Lindström 1964, Müller and Nogami 1971). The basal body was considered by Bengtson (1976, 1983) to be homologous with whole protoconodont and paraconodont elements, although its structure is still not as well known as that of the crown. In investigating the possibility of an evolutionary relationship between protoconodonts and euconodonts, the basal bodies of primitive euconodont elements from Cambrian and lowest Ordovician strata are of particular importance. In these elements, the basal body is usually large relative to the crown and is commonly funnel-shaped with its own basal cavity (Pl. 2.2, figs 1,3). The basal bodies of specimens from the Baltic region are mostly black or dark brown and do not dissolve in hydrochloric acid.

According to Gross (1957), the basal bodies of Ludlovian coniform and ramiform euconodont elements that he investigated are constructed of two separate layers, a basal cone and a cone filling. He considered the basal cone to be more regularly and finely laminated than its filling, and his illustration (Gross 1957, fig. 4) suggests that the lamination of the two layers is discordant. Subsequently, Müller and Nogami (1971) stated that both 'layers' of the basal body are built of the same continuous lamellae, and found that the apparent discordance resulted from a sharp curvature of the lamellae. Hence, they saw no reason to differentiate the basal body into two layers. Furthermore, they also recognized a concordant continuation between the lamellae of the basal body and those of the crown, with the differences between the two being mainly due to the size of the apatite crystallites (see Pietzner *et al*. 1968) and the amount of organic material present. However, my investigations show that in primitive euconodonts the outer part of the basal body (= basal cone) is built of

---

Plate 2.1

Fig. 1—*Sagitta* sp., Recent, north Atlantic, ZPAL C.IV/2.19. Outer view of one incomplete row of teeth, SEM ×460.

Fig. 2—*Sagitta* sp., Recent, north Atlantic, ZPAL C.IV/1.3. Pulp cavity view of two teeth, SEM ×460.

Fig. 3—*Furnishina* sp., Upper Cambrian, Żarnowiec borehole, northern Poland, ZPAL C.IV/1.10. Two juvenile elements fused in natural arrangement, SEM X130.

Fig. 4—*Furnishina* sp., Upper Cambrian, Żarnowiec borehole, northern Poland, ZPAL C.IV/8.1. Two juvenile elements fused in unnatural arrangement, SEM ×90.

Fig. 5a–c—*Furnishina* sp., Upper Cambrian, Uddagården, southern Sweden, ZPAL C.IV/37.3. Polished and etched longitudinal section. 5a, complete specimen, SEM ×230. 5b,c, parts of specimen, SEM ×500.

Fig. 6—*Furnishina* sp., Upper Cambrian, Uddagården, southern Sweden, ZPAL C.IV/36.1. Part of a polished and etched longitudinal section, SEM ×10 000.

Fig. 7—*Furnishina* sp., Upper Cambrian, *Agnostus pisiformis* Zone, Żarnowiec borehole, northern Poland, ZPAL C.IV/T.48. Area of the basal part of a demineralized specimen in ultrathin cross section, TEM ×4 400.

Fig. 8—*Furnishina alata* Szaniawski, Upper Cambrian, *Agnostus pisiformis* Zone, Żarnowiec borehole, northern Poland, ZPAL C.IV/T.53. Area of the cusp of a demineralized specimen in ultrathin cross section, TEM ×22 000.

Fig. 9—*Furnishina* sp., Upper Cambrian, *Agnostus pisiformis* Zone, Żarnowiec borehole, northern Poland, ZPAL C.IV/T.49. Area of the basal part of a specimen that has not been demineralized, ultrathin cross section, TEM ×40 000.

lamellae that differ sufficiently from those secreted subsequently (= cone filling) to be treated as a separate layer (Fig. 2.1B; Pl. 2.2, fig. 1; Pl. 2.3, fig. 1). The basal cone is built of regular, very fine lamellae aligned parallel to the cone surface (Pl. 2.2, fig. 2). These lamellae are composed largely of organic material, with minor amounts of phosphate. The basal filling is constructed of much thicker and less regular lamellae with a much higher mineral content, part of which is evidently of secondary origin. The basal cone is much more resistant to acids, and, as observed by Gross (1957), it is much more commonly preserved (Pl. 2.2, fig. 3).

In demineralized specimens observed with TEM, the organic matter of the basal body is in the form of very fine, dense concentrations, possibly distributed irregularly within the layers (Pl. 2.3, fig. 2). The spaces left by the dissolved phosphate indicate that the phosphatic crystallites were also very fine and probably surrounded the organic concentrations. In the cone filling there are often much larger apatite crystallites of secondary origin (Pl. 2.4, fig. 1b). Additionally, in the cone filling of some specimens, especially of the genus *Cordylodus*, spherulitic structures occur between the laminae (Pl. 2.3, figs. 1,2). Such spherulites were originally noted by Pander (1856) and were illustrated by Müller and Nogami (1971), but their origin is still not understood. The TEM results show that they are composed of

the same material as the regular laminae, arranged in a similar fashion (Pl. 2.3, fig. 2), and thus were probably secreted contemporaneously. The basal bodies of many specimens also often contain secondary pyrite mineralization, which is more extensive in the basal filling than in the cone.

Slow etching of complete primitive euconodont elements usually revealed that their entire crown was coated by a thin translucent layer (Pl. 2.4, fig. 2). This layer extends basally to the surface of the basal body, has a smooth surface, and is resistant to careful treatment with hydrochloric acid. It retains the shape of the crown, after demineralization of the specimen, as long as it is kept in the solvent, but it collapses if dried in air. Critical-point drying, however, enables preservation of this layer and of the underlying organic matter from within the phosphatic lamellae. It has not been established if the layer is of primary origin, and in many cases it has clearly been secondarily phosphatized.

It has long been known that the crown of euconodont elements contain some organic matter (Ellison 1944), and Lindström (1964) observed a 'conodont ghost' after slow, incomplete demineralization of an element in hydrochloric acid. Pietzner *et al.* (1968) identified probable amino acids in the euconodont crown, and illustrated organic matter from a demineralized specimen of *Cordylodus* sp. on a transmission electron micrograph. However,

Plate 2.2
Fig. 1a–b—*Eoconodontus* sp., lower Tremadoc of Estonia, ZPAL C.IV/50.3. 1a, polished and etched longitudinal section, showing the euconodont crown almost completely dissolved and the basal body slightly etched, SEM ×140. 1b, area of same specimen, SEM ×400.
Fig. 2—*Eoconodontus* sp., lower Tremadoc of Estonia, ZPAL C.IV/50.2. Part of a polished and etched longitudinal section, showing the wall of the basal cone, SEM ×8 600.
Fig. 3a–b—*Cordylodus* sp., lower Tremadoc of Estonia, ZPAL C.IV/50.1. 3a, polished and etched oblique section showing the euconodont crown and basal filling dissolved, but the basal cone apparently intact, SEM ×170. 3b, outer surface of the basal cone, SEM ×17 000.
Fig. 4—*Cordylodus* sp., Tremadoc of Estonia, ZPAL C.IV/49.1. Polished and etched outer surface of basal cone, SEM ×12 000.
Fig. 5a–c—*Cordylodus* sp., Tremadoc of Estonia, ZPAL C.IV/50.4. 5a, polished and etched longitudinal section, SEM ×60. 5b, area showing contact between basal body and crown, SEM ×475. 5c, close-up of same area, SEM ×2 100.

**Descriptions of element structures**

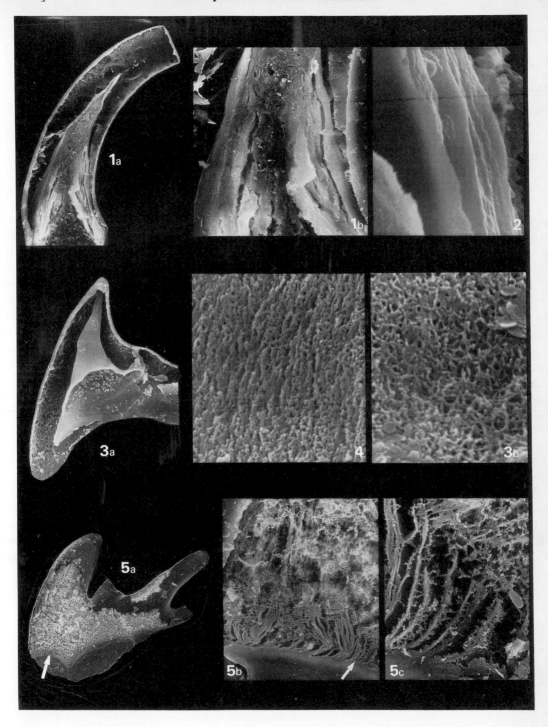

these authors considered that secondary bitu-minous material may invade the phosphatic lamellae and would be very difficult to dif-ferentiate from the primary organic matrix. Fåhræus and Fåhræus-van-Ree (1985, 1987) believe that fixing of the organic matter of conodont elements during demineralization can reveal even the original cellular structure, which may be studied with the optical micro-scope. However, the structure and chemical composition of conodont elements is still not well known, and has not previously been exa-mined with scanning electron microscopy. In the primitive euconodont elements prepared in my investigations, demineralization revealed that under the thin coating of the crown there is an irregular, three-dimensional network con-structed of fibrils (Pl. 2.4, figs 1,2). Holes in the net are of irregular shape and variable size. Although it cannot be unequivocally demon-strated that the network represents primary organic material of the conodont crown, the nearly uniform dimension of the fibrils suggests that this is more likely than a secondary infiltration.

One of the polished and etched thin sections of *Cordylodus* shows that the organic lamellae of the basal body continue for some distance into the crown (Pl. 2.2, fig. 5) where they probably interfinger with its phospahatic lamel-lae. This observation has yet to be confirmed in other specimens. The concordance of crown lamellae with those of the basal body reported by Müller and Nogami (1971) and Lindström and Ziegler (1971) is not visible in my prepar-ations, which suggest that at least some of the initially-secreted lamellae of the basal cone have no continuity with the lamellae of the crown.

## 2.3 DISCUSSION AND CONCLUSIONS

My histological investigations confirm the earlier suggestions (Lindström 1964, Müller and Nogami 1971, Bengtson 1976) that proto-conodont and paraconodont elements are closely similar to the basal body of euconodont elements. The similarities are particularly marked in the case of primitive euconodonts, and may be summarized as follows:

(a) All three element types are composed of organic matter with phosphatic admixtures. The chemical composition of the basal body of euconodont elements evolves from mainly organic in primitive forms to mainly phosphatic in advanced forms, suggesting an origin from elements of organic composition.

(b) All the elements are laminated and show growth towards their bases. In protocodonts and euconodont basal bodies this occurred through addition of lamellae from the cavity side; in paraconodonts the lamellae were added from both the cavity and outer sides.

(c) The organic matter in all the elements is in the form of very fine concentrations. In paraconodonts and euconodont basal bodies it appears to be more regularly arranged, but this may be an effect of preservation.

(d) Phosphate occurs in all elements in the form of very fine equidimensional or slightly elongated crystallites. Transmission electron micrographs suggest that the crystallites are more regularly arranged in paraconodont ele-ments and euconodont basal bodies than in protoconodont elements.

(e) The outer organic coating of proto-conodont and paraconodont elements may correspond to the outer coating of the crown of

---

Plate 2.3
Fig. 1a–c—*Cordylodus* sp., Tremadoc of Estonia, ZPAL C.IV/41.4. 1a, polished and etched longitudinal section, SEM ×90. 1b, close-up, SEM ×330. 1c, close-up, SEM ×660.
Fig. 2a–c—*Cordylodus* sp., Tremadoc of Stora Backor, southern Sweden, ZPAL C.IV/T.63. 2a, close-up of part of cone filling of demineralized specimen in ultrathin cross section, TEM ×3 800. 2b, higher magnification of part of the same area, TEM ×15 000. 2c, another area of the basal body (basal cone?), TEM ×7 500.

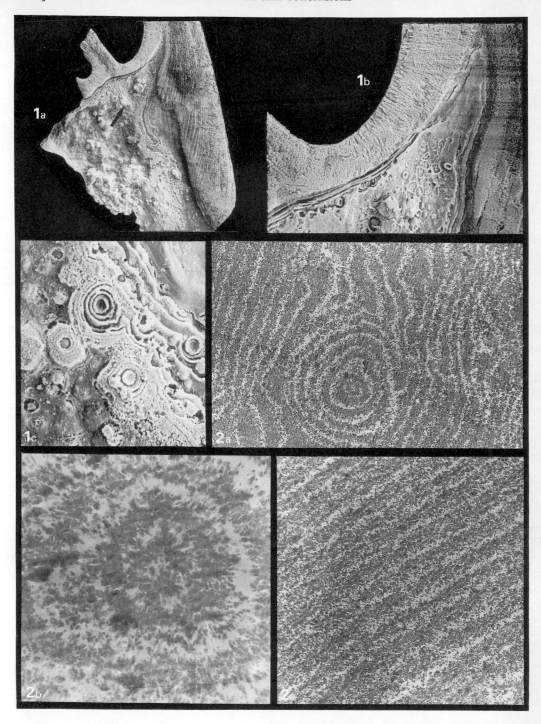

primitive euconodont elements. It is possible that each lamina of the euconodont crown possessed such an organic coating, and that the coatings were continuous with the laminae of the basal body. This is suggested by the preparation illustrated on Pl. 2.2, fig. 5.

The main difference between the element types is that protoconodonts have a well-differentiated inner layer that is not known in paraconodonts or in euconodont basal bodies. In the cone filling of some specimens there is a more finely and regularly laminated inner zone, but the nature of this has yet to be clarified.

My studies support Bengtson's (1976) hypothesis that protoconodont, paraconodont, and euconodont elements are homologous and belong to one evolutionary lineage. The euconodont crown developed later in evolution than the basal body and has no corresponding tissue in protoconodonts or paraconodonts. The ontogenetic development of primitive euconodonts shows the same pattern, with, for example, the basal body of *Cordylodus* developing earlier in ontogeny than its crown (Andres 1981).

Boersma (1972) observed that the basal plate of a Pa element of *Siphonodella* appears to partly transform into crown material during ontogeny. It is possible that the strongly phosphatized tips of some paraconodont elements represent a similar transformation, with the distal parts of early lamellae becoming initial crown (Fig. 2.2).

Taken together, the proposed evolutionary linkage between protoconodonts and euco-nodonts and the evidence of a relationship between protoconodonts and chaetognaths (Szaniawski 1980, 1982, 1983) imply an affinity between conodonts and chaetognaths. A chaetognath affinity was first proposed by Rietschel (1973), but at that time attracted no more attention than the many other hypotheses regarding the relationships of conodonts. The structural evidence documented in this chapter supports this hypothesis, which is also consistent with some other observations:

‘ (a) Conodonts and chaetognaths both show wide global and water-depth distribution. Before there were any published suggestions of conodont–chaetognath affinity, Seddon and Sweet (1971, p. 869) noted that chaetognaths ‘. . . constitute a phylum that is a likely ecologic analogue for conodonts in several ways’.

(b) A functional analogy between some conodont apparatuses and the grasping apparatus of chaetognaths has been noted by many authors (e.g. Rietschel 1973, Repetski and Szaniawski 1981, Aldridge *et al*. 1985, Dzik and Drygant 1986).

(c) Both conodont elements and chaetognath spines display a considerable ability to regenerate.

(d) Both groups are evolutionarily advanced and very distinct from all other members of the Animalia.

(e) The Carboniferous conodont animals described recently (Briggs *et al*. 1983, Aldridge *et al*. 1986) show several similarities to modern chaetognaths (see also Bengtson 1983a,b). Although there are some differences in con-

---

Plate 2.4

Fig. 1a–d—*Eoconodontus* sp., lower Tremadoc of Estonia, ZPAL C.IV/60.1. 1a, polished and etched cross section of the basal part dried by the critical point method, most of the phosphate dissolved, SEM ×230. 1b, close-up of the area denoted by the upper arrow in 1a, showing filling of the basal cone, SEM ×2 300. 1c, close-up of the area denoted by the lower arrow in 1a, showing the basal cone and the crown with organic admixtures, SEM ×1 000. 1d, close-up showing organic material in the crown, SEM ×2 300.

Fig. 2a–b—*Eoconodontus* sp., lower Tremadoc of Estonia, ZPAL C.IV/58.1. 2a, basal part of etched specimen dried by the critical point method, showing remnants of the demineralized crown with fragments of the outer coating and the basal cone visible where the crown has been removed (lower arrow), SEM ×210. 2b, close-up of area indicated by upper arrow in 2a, showing the outer coating and organic material in the crown, SEM ×1 600.

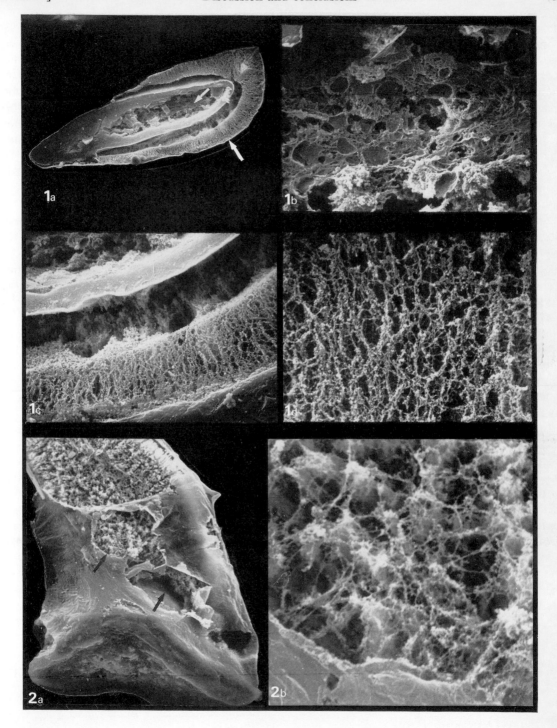

struction between them, these are quite under-standable. Carboniferous conodont elements are so dissimilar to the grasping spines of Recent chaetognaths that we cannot expect them to have belonged to identical animals. The development of the phosphatic crown in conodonts marked the beginning of a new evolutionary line which, in contrast to the chaetognaths, is characterized by rapid evolutionary development and diversification. Conodonts are certainly not chaetognaths *sensu stricto*, a point emphasized by their extinction long ago while chaetognaths remain very well adapted to their environment. However, it is very probable that conodonts and chaetognaths originated from a common ancestor (see also Bengtson 1983a,b). This proposal does not necessarily conflict with hypotheses that conodonts are close to chordates (see Aldridge and Briggs 1986, Aldridge *et al.* 1986). Chaetognaths, like hemichordates, are enterocoelous coelomates, and their affinities are not well understood (Hyman 1959).

Some Cambrian paraconodont elements are similar in shape to the teeth of chaetognaths, rather than their grasping spines. Juvenile paraconodont elements sometimes occur as clusters (Szaniawski 1980, Andres 1981), usually comprising very similar elements in contact at their bases (Pl. 2.1, figs. 3,4). The teeth of Recent chaetognaths occur in similar arrangements (Pl. 2.1, figs. 1,2). It is possible that protoconodont elements correspond to the grasping spines, and paraconodont elements to the teeth of the same, or very similar, animals (see Bengtson 1983b, Szaniawski 1984). Bengtson (1983b) suggested that paraconodont elements could have been the pharyngeal denticles of animals representing an intermediate evolutionary step between Cambrian chaetognaths and conodonts.

## ACKNOWLEDGEMENTS

Dr Viive Viira, Institute of Geology, Estonian Academy of Sciences, Tallinn, provided many well-preserved conodont specimens for histological study. Professor Gisle Fosse, University of Bergen, and Dr Stefan Bengtson, Uppsala University, made available their SEM laboratory facilities. Dr Bengtson, Dr Dietmar Andres, Free University, Berlin, and Dr Jerzy Dzik, Academy of Sciences, Warsaw, read the manuscript and made many valuable comments. The text was edited linguistically by Dr Richard Aldridge, University of Nottingham. I am deeply indebted to all of them.

## REFERENCES

Aldridge, R. J. and Briggs, D. E. G. 1986. Conodonts. In: A. Hoffman and M. H. Nitecki (eds): *Problematic fossil taxa*, Oxford University Press.

Aldridge, R. J., Briggs, D. E. G., Clarkson, E. N. K., and Smith, M. P. 1986. The affinities of conodonts: new evidence from the Carboniferous of Edinburgh, Scotland. *Lethaia* **19** 279–291.

Aldridge, R. J., Briggs, D. E. G., and Smith, M. P. 1985. The structure and function of panderodontid conodont apparatuses. In: R. J. Aldridge, R. L. Austin, and M. P. Smith (eds): *Fourth European Conodont Symposium (ECOS IV), Nottingham 1985, Abstracts*, University of Southampton, 2.

Andres, D. 1981. Bezeichungen zwischen kambrischen Conodonten und Euconodonten (Vorläufige Mitteilung). *Berliner Geowissenschaftliche Abhandlungen* **A32** 19–31.

Bengtson, S. 1976. The structure of some Middle Cambrian conodonts, and the early evolution of conodont structure and function. *Lethaia* **9** 185–206.

Bengtson, S. 1983a. A functional model for the conodont apparatus. *Lethaia* **16** 38.

Bengtson, S. 1983b. The early history of the Conodonta. *Fossils and Strata* **15** 5–19.

Boersma, K. T. 1973. On the basal structure of *Siphonodella cooperi* Hass, 1959 and *Siphonodella lobata* (Branson and Mehl), 1934. *Leidse Geologische Mededelingen* **49** 39–57.

Bone, Q., Ryan, K. P., and Pulsford, A. 1983. The structure and composition of the teeth and grasping spines of chaetognaths. *Journal of the Marine Biological Association of the United Kingdom* **63** 929–939.

Briggs, D. E. G., Clarkson, E. N. K., and Aldridge, R. J. 1983. The conodont animal. *Lethaia* **16** 1–14.

Dzik, J. and Drygant, D. M. 1986. The apparatus of panderodontid conodonts. *Lethaia* **19** 133–141.

Ellison, S. P. Jr 1944. The composition of conodonts. *Journal of Paleontology* **18** 133–140.

Fåhræus, L. E. and Fåhræus-van Ree, G. E. 1985. Histomorphology of soft tissue conodont matrix. In: R. J. Aldridge, R. L. Austin, and M. P. Smith (eds): *Fourth European Conodont Symposium (ECOS IV), Nottingham, 1985, Abstracts*, University of Southampton, 10–11.

Fåhræus, L. E. and Fåhræus-van Ree, G. E. 1987. Soft tissue matrix of decalcified pectiniform elements of *Hindeodella confluens* (Conodonta, Silurim). In: R. J. Aldridge (ed.): *Palaeobiology of Conodonts*, Ellis Horwood, Chichester, Sussex, 105–110.

Gross, W. 1957. Über die Basis der Conodonten. *Paläontologische Zeitschrift* **31** 78–91.

Hyman, L. H. 1959. *The invertebrates: Volume 5, Smaller Coelomate Groups*. McGraw-Hill Book Co., New York, 783 pp.

Lindström, M. 1964. *Conodonts*. Elsevier, Amsterdam, 196 pp.

Lindström, M. and Ziegler, W. 1971. Feinstrukturelle Untersuchungen an Conodonten. 1. Die Überfamilie Panderodontacea. *Geologica et Palaeontologica* **5** 9–33.

Müller, K. J. 1959. Kambrische Conodonten. *Zeitschrift der Deutschen Geologischen Gesellschaft* **111** 434–485.

Müller, K. J. 1962. Supplement to systematics of conodonts. In: R. C. Moore (ed.): *Treatise on Invertebrate Paleontology, Part W, Miscellanea*. Geological Society of America and University of Kansas Press, Lawrence, Kansas. W246–W249.

Müller, K. J. and Andres, D. 1976. Eine Conodontengruppe von *Prooneotodus tenuis* (Müller, 1959) in natürlichen Zusammenhang aus dem Oberen Kambrium von Schweden. *Paläontologische Zeitschrift* **50** 193–200.

Müller, K. J. and Nogami, Y. 1971. Über den Feinbau der Conodonten. *Memoirs of the Faculty of Science, Kyoto University, Series of Geology and Mineralogy* **38** 87 pp.

Müller, K. J. and Nogami, Y. 1972. Growth and function of conodonts. *24th International Geological Congress, Montreal, Section 7*, 20–27.

Pander, C. H. 1856. *Monographie der fossilen Fische des Silurischen Systems der Russisch-Baltischen Gouvernements*, Akademie der Wissenschaften, St. Petersburg 1–91, 7 pls.

Pietzner, H., Vahl, J., Werner, H., and Ziegler, W. 1968. Zur chemischen Zusammensetzung und Mikromorphologie der Conodonten. *Palaeontographica A* **128** 115–152.

Repetski, J. E. and Szaniawski, H. 1981. Paleobiologic interpretation of Cambrian and earliest Ordovician conodont natural assemblages. In: M. E. Taylor (ed.): *Short papers for the second International Symposium on the Cambrian System, 1981, U.S. Geological Survey Open-file Report* 81–743, 169–172.

Rietschel, S. 1973. Zur Deutung der Conodonten. *Natur und Museum* **103** 409–418.

Schram, F. R. 1973. Pseudocoelomates and a nemertine from the Illinois Pennsylvanian. *Journal of Paleontology* **47** 985–989.

Seddon, G. and Sweet, W. C. 1971. An ecologic model for conodonts. *Journal of Paleontology* **45** 869–880.

Szaniawski, H. 1971. New species of Upper Cambrian conodonts from Poland. *Acta Palaeontologica Polonica* **16** 401–413.

Szaniawski, H. 1980. Fused clusters of paraconodonts. In: H. P. Schönlaub (ed.): *Guidebook, abstracts, Second European Conodont Symposium (ECOS II), Abhandlungen der Geologischen Bundesanstalt* **35** 211.

Szaniawski, H. 1982. Chaetognath grasping spines recognized among Cambrian protoconodonts. *Journal of Paleontology* **56** 806–810.

Szaniawski, H. 1983. Structure of protoconodonts. *Fossils and Strata* **15** 21–27.

Szaniawski, H. 1984. Structure and possible origin of paraconodonts. *27th International Geological Congress, Moscow, Abstracts* **9** (supplement 2), 64–65.

# 3

# Re-examination of Silurian conodont clusters from Northern Indiana

R. S. Nicoll and C. B. Rexroad

## ABSTRACT

Examination of fused conodont clusters belonging to the genus *Ozarkodina* from Silurian strata in northern Indiana, including those described by Pollock (1969), demonstrates that these species conform to the general septimembrate apparatus structure described for most middle Palaeozoic conodont species. In this study we have examined original collections and new collections of both fused and discrete elements. Both types of material indicate that Silurian species of *Ozarkodina* possessed septimembrate apparatuses containing 15 elements. The clusters have an anterior complex of discernens elements followed by the contundens elements. Crystal overgrowths obscure the fine detail of some of the clusters, but enough good specimens have been examined to provide an interpretation of the apparatus structure.

## 3.1 INTRODUCTION

Fused conodont clusters were first described from Silurian rocks of northern Indiana (Rexroad and Nicoll 1964) with the recovery of two fused element pairs. Both of these pairs are now assigned to *Ozarkodina eosteinhornensis* (Walliser). Pollock (1969) described an additional 54 conodont clusters from six more localities in northern Indiana. All but five of Pollock's specimens were obtained from the Kokomo Limestone Member of the Salina Formation at the abandoned France Stone Company quarry (Fig. 3.1) at Kenneth, Cass County, Indiana, E1/2 SW1/4 SE1/4 sec. 30, T.27 N., R. 1E.

The present study is based on extensive re-collecting of the quarry site (Fig. 3.2) from which more than 700 additional clusters were recovered. Some of the material is very fragmented, and element identification is difficult. Elements of *Ozarkodina* are relatively thin, and sediment compaction appears to have been responsible for much of the element breakage. Other specimens are so overgrown by secondary phosphate crystals that it is not possible to accurately count or identify the elements in the clusters. A general problem with all studies of cluster material is the obscuring of diagnostic parts of some elements by the stacking of the elements in the cluster, and the more complete clusters of this study all present this problem.

To circumvent some of these problems we have examined both clusters and discrete elements from samples used in this study. Because the conodont fauna is of limited diversity (only five species) we have been able to determine the multielement associations using mor-

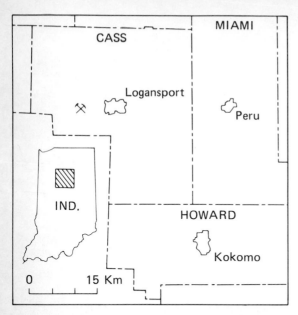

Fig. 3.1—Location of the abandoned France Stone Company quarry, north-central Indiana.

phological criteria in addition to the presence of elements in the clusters. In none of the clusters have we observed any mixing of elements of different species.

The material is reposited in the Indiana University—Indiana Geological Survey collections. Repository numbers (five digits) and locality and sample designations (in parentheses) are included in the plate explanations for all figured specimens.

## 3.2  DISCRETE ELEMENTS

All discrete elements examined can be referred to one of five multielement species: *Belodella* sp., *Dapsilodus?* sp., *Ozarkodina snajdri* (Walliser), *O. eosteinhornensis* (Walliser), and *O.* sp. nov. The discrete element fauna from the cluster locality has been described by Pollock and Rexroad (1973), and our re-collecting has not changed their observations on species distribution. *Ozarkodina eosteinhornensis* is the most abundant species. *O. snajdri* and *O.* sp. nov. are about equally abundant, and the other

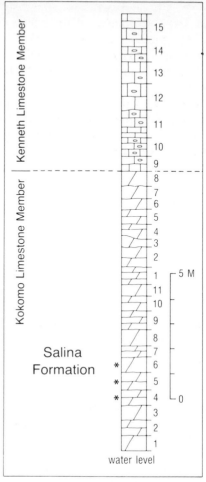

Fig. 3.2—Columnar section showing sample intervals in the Salina Formation from the northeastern face of the abandoned France Stone Company quarry (* indicates samples with clusters).

species are rare. *O. eosteinhornensis* is dominant in the clusters, and *O.* sp. nov. is rare.

Lack of fused clusters and the very low abundance of *Belodella* and *Dapsilodus?* have made analysis of their multielement composition impossible. The composition of the *Ozarkodina* sp. nov. apparatus is relatively easily determined by the robust morphology and round, free standing denticles of the *discernens* elements. It is known from only one partial cluster (Pl. 3.1, fig. 10).

Except for the Pa elements and some Pb elements we have been unable to separate discrete components of the apparatuses of *O. snajdri* and *O. eosteinhornensis*. Examination of the Pa elements of these species (Pollock and Rexroad 1973) indicates a nearly complete gradation of morphology between the two, and if this is the case, it is probable that the discernens elements are not sufficiently differentiated to separate them. Thus clusters of these two species generally cannot be distinguished unless they contain Pa elements.

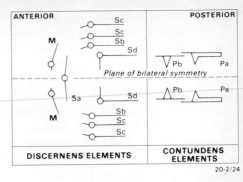

Fig. 3.3—Plan showing arrangement of discernens and contundens elements in *Ozarkodina*.

## 3.3 CLUSTERS: GENERAL CONSIDERATIONS

The clusters range from those that contain a complete apparatus structure of 15 elements to those comprising only a single pair of elements. The positioning of elements in the clusters indicates that some must have had only a minimal disturbance of the elements from their functional position, while others show complete disruption of the element sequence.

Uyeno (1982) and Nicoll (1985) suggested that the apparatus of *Ozarkodina* contains seven element types. This is confirmed by our material which shows that the distribution of element types in the *Ozarkodina* animal (Fig. 3.3) must have been as follows: M = 2, Sa = 1, Sc = 4, Sb = 2, Sd = 2, Pb = 2, and Pa = 2. This is the same ratio of element frequency as found in the *Polygnathus* clusters described by Nicoll (1985). Those clusters thought by Pollock (1969) to contain two or three Sa elements and that with three Pb elements identified have been re-examined. In all cases we recognize only a single Sa element, and the other elements that Pollock had thought to be Sa elements are Sb elements. In the crush of elements in some clusters it can be very difficult to distinguish between these element types. Pollock's example with multiple Pb elements represents a normal cluster; the mis-identified elements are broken Sc and Sb or Sd elements.

In the better preserved, more complete, clusters there appear to be three basic forms of apparatus preservation: piled up, opened out, or chaotic. We think that the first two of these are a reflection of the position of the animal after its death, and that clusters of these types have been subjected to only minimal disturbance (Fig. 3.4). Chaotic preservation reflects disturbance of the elements either by biological activity or current action.

In cross-section the conodont animal was most probably laterally compressed (Briggs *et al*. 1983). Thus, if after death the animal settled on a flat bottom, it will usually have come to rest on its side. Sometimes, however, the animal will have settled on its dorsal or ventral surface. As the soft tissue of the animal decayed the elements will gradually have collapsed onto the surface of the sediment. If the animal was resting on its dorsal or ventral

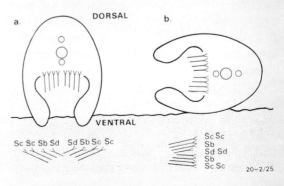

Fig. 3.4—Cross-section of dead conodont animal showing method of formation of (a) opened out and (b) piled clusters.

surface, and if the elements were in a position similar to that proposed in the reconstruction of the animal by Nicoll (1985), then the discernens elements when not constrained by tissue would have tended to fall away from the centre line of the apparatus (Fig. 3.4a). The contundens elements, being oriented transversely, would have remained in their functional positions. This would result in clusters preserved as those shown on Pl. 3.2, figs 1, 3, 4, and by Pollock (1969, Pl. 112, figs 1, 2, 4,).

When the animal came to rest on its side, the discernens elements would have tended to compress laterally into a pile similar to clusters shown on Pl. 3.2, figs 2, 6, 7, with all of the basal cavities pointing in one direction (Fig. 3.4b). The contundens elements would possibly have moved towards the opposing element but would also have fallen laterally to the sediment surface. The M and Sa elements, with their long axes transverse to the axis of the animal, would have fallen into a variety of positions.

One other explanation of element relationships is resolved by the cluster shown on Pl. 3.2, fig. 2. This piled cluster was recovered with the two Pa elements and one Pb element preserved in the process of being peeled away from the rest of the elements. This explains the relatively common recovery of clusters of Pa and Pb elements that appear to have the Pb elements located behind the Pa elements (Pollock 1969, pl. 111, fig. 11).

These theoretical interpretations of postmortem element movement are consistent with the observed distributions in the recovered clusters. Piled clusters are more common than opened out clusters, and the location of M and Sa elements in piled clusters is highly variable. The majority of clusters are chaotic or incomplete with only a few elements preserved. Because preservation of the elements as a cluster depends on the elements being in contact with adjoining elements, it would only take a small amount of disturbance by scavengers or currents to separate the elements and prevent the formation of clusters. The clusters are also very fragile, and many may be broken during processing to remove them from the rock.

## 3.4 CLUSTERS: EXAMPLES

The arrangement of discernens elements is best seen in the opened out clusters. The clearest example is shown on Pl. 3.2, fig. 1, and has a pair of Sd elements located on either side of the central line of the apparatus. Outside and underneath the Sd elements are the two Sb elements followed by two pairs of Sc elements. The single Sa element is located at the anterior end of the apparatus on the under surface; only part of the element can be seen sticking out from beneath the other elements. A single M element is preserved on the left side, and a pair of Pb elements is found near the posterior end of the cluster. This accounts for 12 of the 15 elements that would be found in a complete cluster.

One of the more common cluster types is composed of a pair of Pa elements. In all examples examined, the two elements are of the same size, are overlapped with the denticles of one element almost touching the expanding lip of the basal cavity of the other

---

Plate 3.1—All ×65.
Five digit number is IU–IGS repository number, [1, 5L] indicates locality 1, sample 5L.
Figs. 1–6—*Ozarkodina eosteinhornensis* (Walliser 1964), discrete discernens elements, all [1, 5L]. 1, Sa element, 16821, posterior view. 2, Sd element, 16822, lateral view. 3, Sb element, 16823, lateral view. 4, M element, 16824, posterior view. 5, Sc element, 16825, lateral view. 6, Sc element, 16826, lateral view.
Fig. 7—stereo photomicrographs of Pa element pair, 16827, [1, 4], lateral view.
Figs 8–9—*Ozarkodina snajdri* (Walliser 1964) stereo photomicrographs. 8, Pa element pair, 16828, [1, 5L], lateral view, 9, Pa element pair, 16829, [1, 5L], lateral view.
Fig. 10—*Ozarkodina* sp. nov., stereo photomicrographs of cluster of discernens elements, 16830, [1, 5L], lateral view, includes one Sb and two Sc elements.

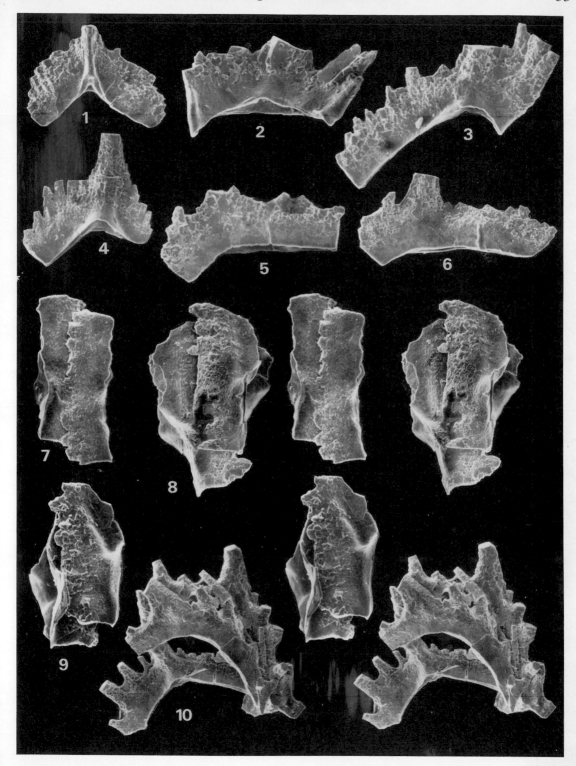

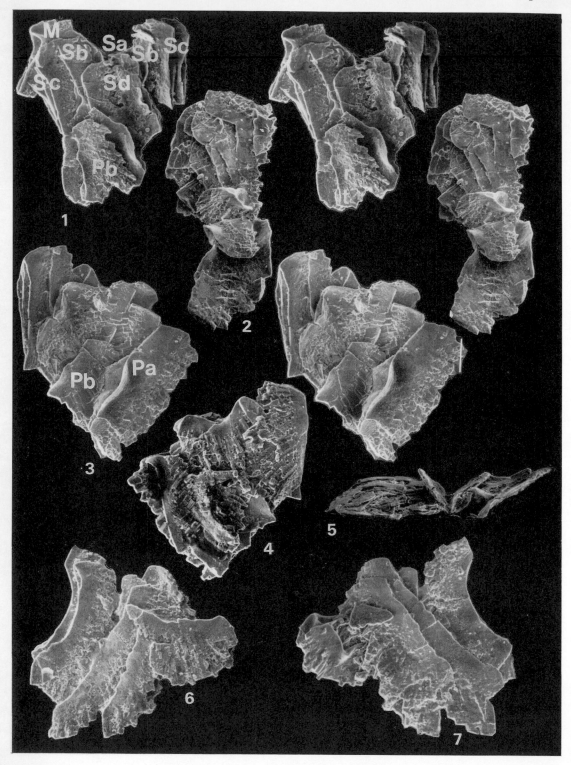

element, and have their anterior ends pointing in the same direction (Pl. 3.1, figs 7–9).

A number of clusters also contain one pair each of Pa and Pb elements. The Pb elements may be located behind, beside, or in front of the Pa elements. In some of the examples studied the Pa element pair is much larger than the Pb element pair. In other clusters the pairs are of similar size, and in the discrete elements the largest Pb elements are as large as the largest Pa elements. It is possible that in these species the Pa elements grew faster than the Pb elements, and only in mature or gerontic forms are the Pa and Pb elements of similar size.

Clusters composed of a pair of Sc elements are relatively common. In all cases the pair consists of either two right or two left elements. In such cases it is impossible to distinguish morphologically the inner and outer elements. Clusters with two Sc elements and an Sb element on the inner side are present but not common. However, larger clusters are present that contain two pairs of Sc elements, a pair of Sb elements on the inner sides, and a single Sa element in the centre of the group (Pollock 1969, pl. 112, figs. 1,2).

## 3.5  COMPARISON WITH PREVIOUS STUDIES OF *OZARKODINA*

The basic apparatus structure of *Ozarkodina* was outlined by Walliser (1964) when he rec-ognized Apparat H (*O. excavata* (Branson and Mehl)) and Apparat J (*O. steinhornensis* (Ziegler)), Jeppsson (1969, 1971, 1974) refined and elaborated the multielement com-position of several species of *Ozarkodina* (*Hindeodella*). Both Walliser and Jeppsson thought that the apparatus consisted of six element types, neither recognizing the Sd ele-ment.

Mashkova (1972) described a bedding plane assemblage of *Ozarkodina steinhornensis* which she interpreted as consisting of 13 ele-ments with the Sa element missing. With refer-ence to Mashkova's specimen (1972, fig. 2, pl. 1) we would interpret the assemblage as follows: M elements—12, 13; Sa ele-ment—missing; Sc elements—8,9,10,11; Sb element—anterior part of 5; Sd element—6; Pb elements 1,2; Pa elements—3,4. The post-erior part of element 5 and element 7 are the broken posterior parts of the left Sb and Sd elements. Elements 5 (anterior), 6,8,10 and 12 are right side elements and elements 7,5 (post-erior), 9,11 and 13 are left side elements. Assuming that the Sa element was present but not recovered, the multielement structure of *O. steinhornensis* is identical to that of *O. eosteinhornensis*.

Uyeno (1982) recognized that *O. brevis* (Bischoff and Ziegler) had a septimembrate apparatus. This was confirmed by Nicoll (1985) from a study of fused clusters and dis-crete elements.

---

Plate 3.2—All ×65.

Figs 1–5—*Ozarkodina eosteinhornensis* (Walliser 1964).

Fig. 1—Stereo photomicrographs of element cluster, 16831, [1, 5L]; inner lateral view showing 12 elements in opened out arrangement: M (1), Sa (1), Sc (4), Sb (2), Sd (2), Pb (2).

Figs. 2,5—Cluster 16832, [1, 4]. 2, stereo photomicrographs of lateral view, showing 14 elements in piled arrangement: M (1), Sa (1), Sc (4), Sb (2), Sd (2), Pb (2), Pa (2); basal cavities of all discernens elements oriented in same direction (left) but those of contundens elements directed away from opposing element; contundens elements show evidence of having been partially peeled back, both Pa elements and one Pb element have been flipped posteriorly but are still attached to the rest of the cluster. 5, basal view.

Figs. 3,4—Cluster 16833, [1, 5L]. 3, stereo photomicrographs of inner lateral view. 4, outer lateral view. Cluster includes 13 elements in opened out arrangement: M (1), Sa (1), Sc (4), Sb (2), Sd (2), Pb (2), Pa (1).

Figs. 6,7—*Ozarkodina* cf. *O. eosteinhornensis* (Walliser 1964), lateral views of cluster 16834, [1, 5U], showing 10 elements in piled arrangement: M (2), Sa (1), Sc (4), Sb (1), Sd (2).

## 3.6 APPARATUS STRUCTURE: *OZARKODINA* COMPARED WITH *POLYGNATHUS*

The element abundance and distribution pattern of *Ozarkodina* is broadly similar to that found in *Polygnathus* clusters (Nicoll 1985) with 15 elements of seven element types. The *Ozarkodina* clusters show the same element sequence as in *Polygnathus* with the M elements anterior and the Pa elements posterior (Fig. 3.5). However, the relationship of the S elements appears to be somewhat different from that observed in *Polygnathus*. In *Ozarkodina* the Sd elements are located inside the Sb elements as an integral part of a compact discernens element complex (Pl. 3.2, figs 1,3). In *Polygnathus* the Sd elements are thought to be located behind the major grouping of discernens elements (Nicoll 1985). These structural differences between *Ozarkodina* and *Polygnathus* apparatuses are probably related to the presence or absence of a posterior process on the Sa element. *Polygnathus*, with a posterior process on the Sa element, does not have space for the Sd elements, with their inwardly directed lateral processes, inside the Sb elements. *Ozarkodina*

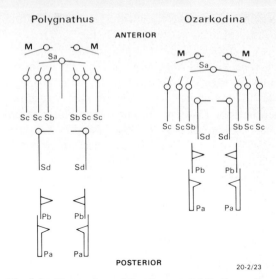

Fig. 3.5—Comparison of the element distribution patterns in *Ozarkodina* and *Polygnathus*.

species that lack a posterior process on the Sa element can have the Sd element located in a position that overlaps the Sb element. This positional relationship of the Sb and Sd elements would result in a more compact grouping of the discernens elements and also allow the contundens elements to be located closer to them. This may be the reason why in *Ozarkodina* the contundens elements are frequently

---

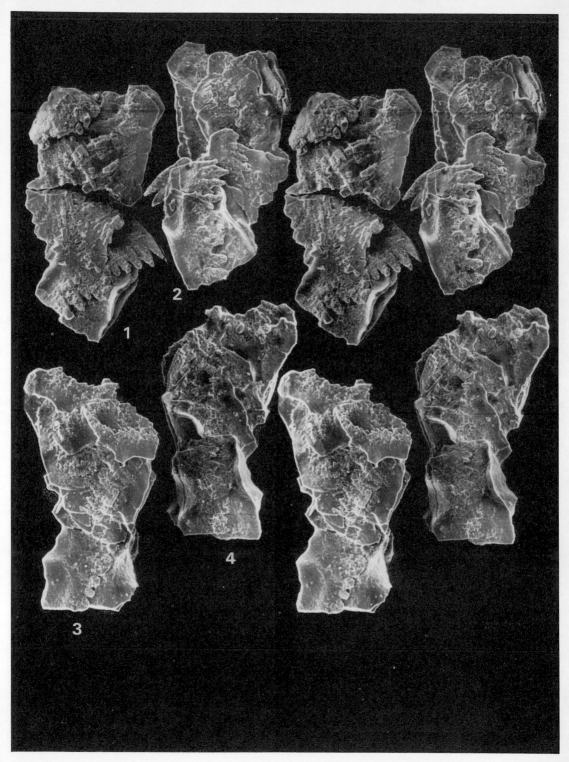

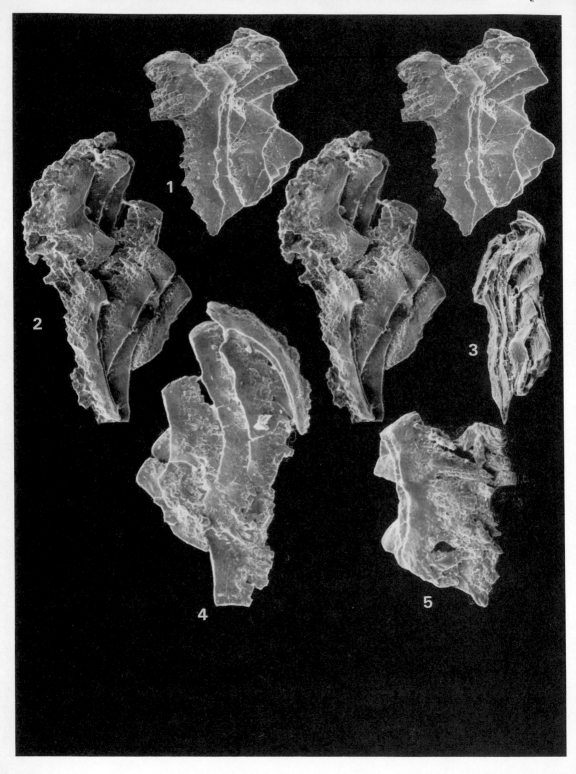

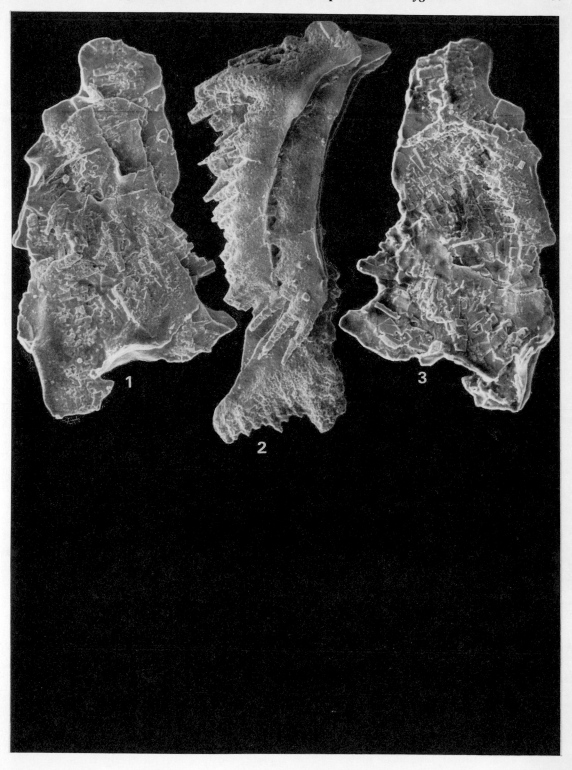

recovered attached to discernens elements in nearly complete clusters, while complete clusters are rare in *Polygnathus*.

Comparison of the preservation of clusters in this study with those described by Nicoll (1983, 1985) shows that complete clusters are more abundant in the Indiana material, while the preservation of element relationships is better in the Australian collection. Except in the best-preserved clusters of this study, the elements appear distorted and out of their presumed functional positions. The lack of a posterior process on the Sa element may be, in part, responsible for this, because there was no element with the lateral support to hold the elements in a stable configuration.

Unlike the clusters of *Polygnathus* examined by Nicoll (1985), in almost all of the clusters of *Ozarkodina* with Pa and/or Pb elements preserved, these are located so that the oral surfaces overlap. We believe this is an indication that in life the elements functioned in a similar position.

### 3.7   MISSING BASAL PLATES

Despite the excellent preservation of conodont elements in the clusters, only the crowns are preserved, and the basal plates are missing. Element morphology and the preservation of basal plates in congeneric material indicate that basal plates must have been present when the elements functioned in the animal.

Cementation of the clusters took place in the sediment some time after burial. Any mechanical disturbance of the conodont elements after the death of the animal and before cementation would have separated the elements, and the clusters could not have formed. This applies to both current action and possible scavenging by multicellular organisms.

Bacterial attack on the remains of the conodont animal, especially on the organic material within the phosphatic structure of the basal plate, could have slightly increased the acidity of the area around the basal plates. These

might thus have been partially, or completely, dissolved. This dissolved material is possibly the source of the phosphate that binds the elements of the clusters together.

Another possibility is that fluids moving through the pore spaces of the rock may have been slightly acidic, and the more open porous structure of the basal plates would have been more easily dissolved than the crowns. An additional factor may be that the basal plates of some genera or species were less mineralized than in other species. Thus, *Icriodus*, which frequently retains basal plate material, may have more mineralization than *Ozarkodina*, which rarely has preserved basal plates.

From our material we can reach no conclusion about the fate of the missing basal plates. We suggest this as an interesting area of investigation.

### 3.8   CONCLUSIONS

The clusters examined in this study represent the preserved skeletal remains of conodont animals. Each cluster may contain all, or only a few, of the elements originally found in the mouth area of the animal. From any one animal only those elements that are in contact with adjacent elements after they have been buried by sediment are preserved in the clusters, and the remainder will be recovered as discrete specimens. The elements of the cluster are held together by secondary phosphate that has been precipitated at some time after their incorporation in the sediment.

None of the clusters shows any evidence of mixing of elements from more than one animal. This we judge from the size similarity of element pairs in the clusters and because we do not find clusters containing more than the expected numbers of each element type.

The apparatus structure of *Ozarkodina* is septimembrate and contains 15 elements. The apparatus is more compact than the apparatus of *Polygnathus*; and complete, or nearly complete, clusters are relatively common in our collections. The relationship of the elements in some of the clusters can be used to interpret the final resting position of the conodont animal on the sea bottom after death.

## ACKNOWLEDGEMENTS

SEM photography and plate preparation were the work of Arthur T. Wilson (Bureau of Mineral Resources, Canberra). Publication is authorized by the State Geologist, Indiana Geological Survey, and the Director, Australian Bureau of Mineral Resources, Geology and Geophysics.

## REFERENCES

Briggs, D. E. G., Clarkson, E. N. K., and Aldridge, R. J. 1983. The conodont animal. *Lethaia* **16** 1–14.

Jeppsson, L. 1969. Notes on some Upper Silurian multielement conodonts. *Geologiska Föreningens i Stockholm Förhandlingar* **91** 12–24.

Jeppsson, L. 1971. Element arrangement in conodont apparatuses of *Hindeodella* type and in similar forms. *Lethaia* **4** 101–123.

Jeppsson, L. 1974. Aspects of Late Silurian conodonts. *Fossils and Strata* **6** 1–54.

Mashkova, T. V. 1972. *Ozarkodina steinhornensis* (Ziegler) Apparatus, its Conodonts and Biozone. *Geologica et Palaeontologica* **SB1** 81–90.

Nicoll, R. S. 1983. Multielement composition of the conodont *Icriodus expansus* Branson and Mehl from the Upper Devonian of the Canning Basin, Western Australia. *BMR Journal of Australian Geology and Geophysics* **7** 187–213.

Nicoll, R. S. 1985. Multielement composition of the conodont species *Polygnathus xylus xylus* Stauffer, 1940 and *Ozarkodina brevis* (Bischoff and Ziegler, 1957) from the Upper Devonian of the Canning Basin, Western Australia. *BMR Journal of Australian Geology and Geophysics* **9** 133–147.

Pollock, C. A. 1969. Fused Silurian conodont clusters from Indiana. *Journal of Paleontology* **43** 929–935.

Pollock, C. A. and Rexroad, C. B. 1973. Conodonts from the Salina Formation and the upper part of the Wabash Formation (Silurian) in north-central Indiana. *Geologica et Palaeontologica* **7** 77–92.

Rexroad, C. B. and Nicoll, R. S. 1964. A Silurian conodont with tetanus? *Journal of Paleontology* **38** 771–773.

Uyeno, T. T. 1982. Systematic conodont paleontology. In: Norris, A. W., Uyeno, T. T., and McCabe, H. R. Devonian rocks of the Lake Winnipegosis—Lake Manitoba outcrop belt, Manitoba. *Geological Survey of Canada, Memoir* **392** 280 pp.

Walliser, O. H. 1964. Conodonten des Silurs. *Abhandlungen des Hessischen Landesamtes für Bodenforschung* **41** 1–106.

# 4

# The architecture and function of Carboniferous polygnathacean conodont apparatuses

R. J. Aldridge, M. P. Smith, R. D. Norby, and D. E. G. Briggs

## ABSTRACT

Bedding plane assemblages of Carboniferous polygnathaceans display a limited number of recurrent patterns. The most common is an arrangement in which the ramiform S and M elements form a sub-parallel cluster set obliquely to parallel pairs of pectiniform Pa and Pb elements. This pattern is evident in the conodont animals from Granton, Scotland, and is herein termed the 'standard pattern'. Other repeated arrangements are designated 'parallel', 'perpendicular', and 'linear'. Assemblages of all these types can be produced by collapse and flattening of the same three-dimensional structure. In this architecture, the elements are arranged about a dorso-ventral plane of symmetry, with the ramiform elements at the anterior and set at a steep angle to the long axis of the animal; the Pb and Pa pairs follow posteriorly, lying vertically and nearly normal to the long axis of the trunk. An integrated functional system may be postulated for the apparatus, with the anterior elements grasping food to be processed by a shearing and grinding action of the pectiniform elements.

## 4.1 INTRODUCTION

The superfamily Polygnathacea encompasses those conodonts whose apparatuses include a Pa element which can be derived, directly or indirectly, from the Pa element of *Ozarkodina* (Klapper, In: Robison 1981, p. W157). The complete apparatus is normally seximembrate or septimembrate with the ramiform elements multidenticulate. There is ample evidence of the multielement composition of members of this group, as the vast majority of known bedding plane assemblages and many clusters are of polygnathacean genera. Additional important information is provided by the first complete conodont animal, from the Carboniferous of Scotland, which is a cavusgnathid polygnathacean (Briggs *et al.* 1983). Hence, much of the discussion of apparatus structure and function has centred on the polygnathacean plan.

The elemental composition of the polygnathaceans may be exemplified by the apparatus of *Gnathodus bilineatus* (Roundy), which occurs as bedding plane assemblages in the Tyler Formation of central Montana (Norby 1976). This apparatus is septimembrate (Fig. 4.1), comprising one pair each of platform (Pa), angulate (Pb), and dolabrate ramiform (M) elements, and a set of ramiform S elements. Within the group of S elements are four pairs with bipennate morphology; two of these pairs are designated Sc elements, the other two Sb and Sd respectively. There is a single, bilaterally symmetrical, alate Sa element. In some related species and genera an Sd element is not differentiated, and there may be two pairs each of Sb and Sc elements or, possibly, four pairs of Sc elements.

Although the multielement composition of several polygnathacean genera is well understood, it has proved difficult to use the bedding plane assemblages to interpret the original three-dimensional arrangement of the elements. The development of ideas on apparatus architecture is outlined in Chapter 1 of this volume, where various skeletal plans that have been proposed to date are illustrated (Aldridge 1987, fig. 1.5). Many assemblages are essentially two-dimensional arrays of elements resulting from the collapse, after death, of the three-dimensional apparatus. Others may be affected by decay processes and by the activities of scavengers and burrowers. Some originate as faeces or regurgitated gastric residues, but these can normally be recognized by their disorganized nature. Examination of collections of well-preserved bedding plane assemblages, however, reveals several that show consistent arrangements of elements which may be interpreted as natural flattened configurations. These repeated patterns are limited in number; here we designate the most common as the 'standard' pattern, others are termed 'parallel', 'perpendicular', and 'linear'. The configuration of elements reflects the orientation of the apparatus relative to the plane of flattening, modified by any rotation of elements that occurred during collapse or as a result of muscular contraction on the animal's death. Any interpretative model of the three-dimensional architecture of apparatuses must account for these patterns through a simple process of death and collapse/compaction.

Our terms for the common arrangements are descriptive, and differ from those used previously by Collinson *et al*. (1972), who interpreted the patterns as derived from an originally linear array (*op cit*., fig. 10A). Norby (1976, 1979) subsequently suggested that the 'simple rotation' pattern (Collinson *et al*. 1972, fig. 10E) might more closely reflect the three-dimensional architecture of the apparatus. It should be noted that the illustrations of com-

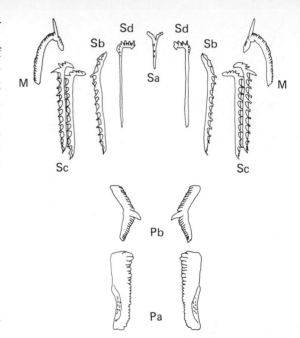

Fig. 4.1—The multielement composition of a Carboniferous polygnathacean apparatus; schematic diagram based on *Gnathodus bilineatus*.

mon arrangements given by Collinson *et al*. (1972, fig. 10) are idealized and are not all based on actual assemblages.

## 4.2   THE STANDARD PATTERN

Assemblages of this type (Fig. 4.2A) show the elements segregated into three sets: a pair of parallel Pa elements, with their denticulated free blades in opposition, a pair of Pb elements, again parallel with anterior ends adjacent and denticles opposed, and a group of ramiform M and S elements with their posterior processes in sub-parallel alignment. The Pa and Pb pairs are parallel to each other, but the ramiform group is oriented at an angle, with the distal tips of the posterior processes closest to the Pb elements. The denticulated surfaces of the ramiform elements are not usually opposed but commonly face away from the Pb pair, with the M elements situated most distally. The apparatuses in the Granton ani-

A

B

Fig. 4.2—The standard pattern; A, bedding plane assemblage from Bailey Falls, La Salle County, Illinois, specimen no. X-6377, University of Illinois, ×25; B, head region of counterpart of the first specimen of the conodont animal, from the Granton shrimp bed, Edinburgh, specimen no. IGSE 13822, British Geological Survey Edinburgh, ×30.

mals (Briggs *et al*. 1983, Aldridge *et al*. 1986; Fig. 4.2B) display typical standard patterns, and reveal that the ramiform group is at the anterior. The pectiniform pairs are aligned with their long axes near normal to the longitudinal axis of the animal. The differing patterns of denticulation on the posterior processes of the M and S elements result in the denticles of the former pointing forwards, whereas those of the latter are directed backwards.

Polygnathacean assemblages displaying the standard pattern, or slight modifications of it, have been widely illustrated (e.g. Du Bois 1943, pl. 25, figs. 3, 17, 21, Schmidt and Müller 1964, figs. 6,7, Mashkova 1972, pl. 1). Collinson *et al*. (1972) included assemblages of this

type within their 'simple rotation' category, to which Avcin (1974) assigned 49 (48%) of his ordered assemblages from Illinois. Norby (1976) recognized eight specimens of this type and a further eight that were broadly similar and could be ascribed to 'rotation' in his collections from Montana.

The most common variation on the standard theme involves a departure from the rectilinear arrangement of the pectiniform pairs (Fig. 4.3). In this 'offset standard' pattern, the pairs of P elements are offset from each other, and the members of each pair are also displaced. Specimens displaying the 'offset standard' configuration have previously been illustrated by Rhodes (1952, pl. 126, fig. 9) and Schmidt and Müller (1964, figs. 1, 5).

Fig. 4.3—The offset standard pattern: bedding plane assemblage from Bailey Falls, La Salle County, Illinois, specimen no. 5830/016, University of Nottingham, ×17.

## 4.3 THE PARALLEL PATTERN

In this arrangement, all the elements lie parallel to each other (Fig. 4.4). They are characteristically crowded together and commonly superimposed, often obscuring the orientation of the denticulated surfaces of the ramiform elements. The pectiniform pairs normally display opposed denticulate surfaces, but may be offset. Examples include those of Du Bois (1943, pl. 25, figs. 4, 5, 12, 13). Collinson *et al.* (1972) referred to such arrangements as the result of 'simple contraction', and Avcin

(1974) found that 32 (31%) of his ordered assemblages belonged to this category. Norby (1976) considered that 50 of his assemblages 'may be due to contraction', but not all of these match the features of the parallel pattern as we understand it. It is possible that some examples interpreted as 'simple contraction' may be faecal (Norby 1976, p. 50).

## 4.4 THE PERPENDICULAR PATTERN

This uncommon arrangement is typified by the specimen in Fig. 4.5. The pectiniform elements are arranged in longitudinal pairs, with denticulated surfaces opposed. Their posterior ends are directed towards the ramiform set, in which the elements are aligned at a high angle (but not fully perpendicular) to the axes of the pectiniforms. There are few other known assemblages of this type and all show differences in detail.

Fig. 4.5—The perpendicular pattern: bedding plane assemblage from Bailey Falls, La Salle County, Illinois, specimen no. 57–P–170II, Illinois State Geological Survey, ×25.

## 4.5 THE LINEAR PATTERN

In this pattern the three sets of elements are arranged in sequence about a line of bilateral symmetry, across which the denticulated surfaces of element-pairs are opposed.

Fig. 4.4—The parallel pattern: bedding plane assemblage from Bailey Falls, La Salle County, Illinois, specimen no. 57P–38, Illinois State Geological Survey, ×20.

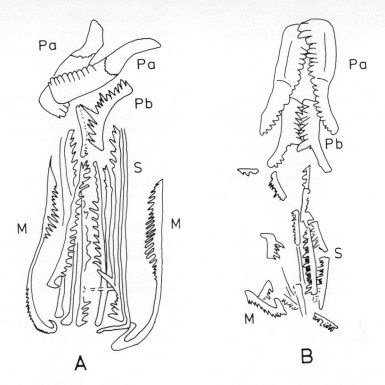

Fig. 4.6—The linear pattern: A, linear 1, bedding plane assemblage from the *bisulcatum*—Kieselschiefer,
Hemer, Westphalia, redrawn after Schmidt 1934, fig. 5b, ×13; B, linear 2, bedding plane assemblage from
Bailey Falls, La Salle County, Illinois, specimen no. X-1480, University of Illinois, ×19.

There are two distinct types of linear arrangement. The first (linear 1) is exemplified by a specimen illustrated by Schmidt (1934, fig. 5a, b; pl. 6, fig. 1). All the elements are aligned so that their posterior processes point in the same direction (Fig. 4.6A). If conventional element orientations were used as an indication of the orientation of the apparatus, the ramiform elements would be at the anterior, followed by the Pb elements and, finally, the Pa elements. The second type of linear array (linear 2) is that shown by the specimen figured by Du Bois (1943, pl. 25, fig. 14, see also Fig. 4.6B) and subsequently re-illustrated several times (e.g. Rhodes 1952, pl. 126, fig. 11, Sweet 1985, fig. 1, von Bitter and Merrill 1985, pl. 1, fig. 1, Aldridge 1987, fig. 1.6 this volume). The anterior–posterior orientation of the element pairs, as far as can be determined, is again concordant throughout

the assemblage, but, in contrast to linear 1, conventional orientation would place the Pa elements at the anterior, followed by the Pb elements. The ramiform group is somewhat disrupted, but is commonly illustrated as following the Pb pair with the posterior processes directed backwards (e.g. Rhodes 1952, fig. 2).

Linear arrays of both types are rare, but have attracted particular attention because of their bilateral symmetry. Many workers have stated or implied that such patterns most closely reflect the original arrangement, and have used the resultant concept of the apparatus architecture to develop functional models (e.g. Schmidt 1934, Hitchings and Ramsay 1978, Nicoll 1985). Further, in their categorization of assemblage patterns, Collinson *et al.* (1972) considered all other configurations to be displacements from an originally linear architecture.

## 4.6 THE ARCHITECTURE OF POLYGNATHACEAN APPARATUSES

Previous workers have found it difficult to reconcile the variety of bedding plane assemblage patterns with a single, three-dimensional apparatus structure. Collinson *et al.* (1972, p. 24). for example, considered that departures from linear arrangements 'can in most cases be explained by muscle relaxation–contraction effects or by compression–decomposition effects that translated elements after death from their life position'. Particular difficulty has been caused by the disparity in the two linear patterns, described herein as linear 1 and linear 2. Jeppsson (1971) concluded that most known assemblages and clusters supported the linear architecture deduced from assemblages of linear 1 type by Schmidt (1934; see also Aldridge 1987, fig. 1.5A), but recognized that the specimen of linear 2 pattern figured by Du Bois (1943, pl. 25, fig. 14) was incompatible. Jeppsson (1971, p. 121), however, noted that elements in several assemblages have been 'turned and twisted', and suggested that parts of Du Bois' apparatus may have been dislocated during post-mortem disintegration. He even tentatively proposed that, since the elements are not uniform in size, this assemblage might comprise the remains of more than one individual.

A major constraint on previous attempts to reconstruct apparatus architecture has been the assumption that the structure was linear, with the long axes of the elements aligned parallel to the axis of the conodont animal. The discovery of complete animals from Edinburgh has shown that this was not necessarily the case. The assemblages in these specimens are not arranged linearly, but in a standard pattern, with the ramiform elements oblique and the pectiniform elements near normal to the trunk of each animal (Fig. 4.2B). This pattern represents a natural death orientation, incorporating the effects of any post-mortem muscular contraction and subsequent collapse of the head of the animal onto the bedding plane.

Collinson *et al.* (1972) invoked 'rotation' of elements to produce assemblages of this type, but we show below that linear and standard patterns can both be produced by collapse and flattening of the same three-dimensional structure. Indeed, not only is the standard pattern the most common of the bedding plane arrangements, but it also shows transition to all the other major types, suggesting that it does reflect the original geometry of the apparatus.

Our approach to unravelling the architecture of the polygnathacean apparatus has been to use the standard pattern as a template for a three-dimensional model (Fig. 4.7) with which we could attempt to explain all the main categories of assemblage simply by flattening and without invoking post-mortem contraction or displacement of elements. Our model was based primarily on the apparatus preserved in the first Granton animal (Fig. 4.2B). The Pa and Pb element pairs are positioned transverse to the long axis of the apparatus, whereas the M and S elements are at an oblique angle calculated from the average orientation of the posterior processes of the ramiform elements in the Granton animal. Within this ramiform set a posterior process with slightly more even denticulation can be discriminated, positioned at a slightly higher angle to the axis than the remainder. This is probably the Sa element, which can also be recognized at a similar orientation in other bedding plane assemblages. We have accordingly followed this positioning in our model. Similarly, we have used evidence from bedding plane assemblages and from clusters to place the M elements outside and slightly anterior to the Sc elements (see also Jeppsson 1971, Nicoll 1985). There is ample evidence from element pairing and bedding plane assemblages that the conodont apparatus was bilaterally symmetrical, so our model includes a dorso-ventral plane of symmetry. As the full suite of elements in the Granton animal is not clearly discernible, we have used the better-known apparatus of *Gnathodus bilineatus* (see Norby 1976) as a basis for modelling the individual elements.

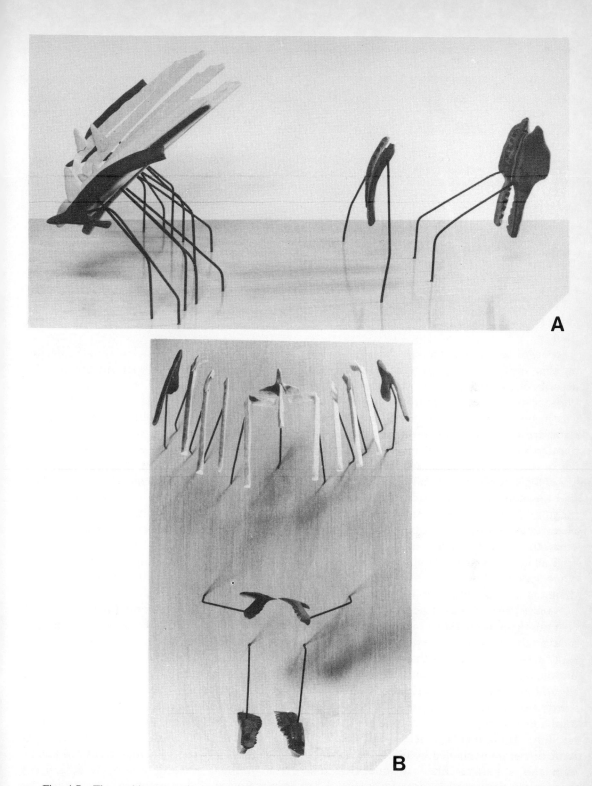

Fig. 4.7—The architecture of a polygnathacean apparatus, reconstructed using modelling clay. The ramiform elements are more widely spaced than in nature to emphasize their relative positions. A, lateral view, anterior to the left; B, view from above the model. In succeeding illustrations, the photographs have been retouched to remove the background and supporting pins.

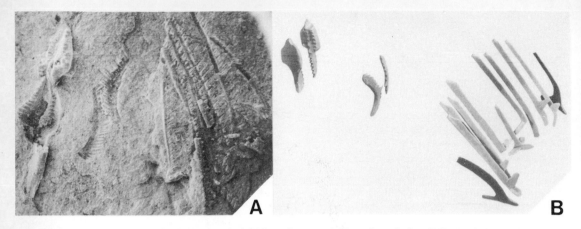

Fig. 4.8—The offset standard pattern: A, bedding plane assemblage from Bailey Falls, La Salle County, Illinois, specimen no. 57P–72I, Illinois State Geological Survey, ×15; B, oblique view of model.

In order to examine the effects of flattening on this structure we employed the approach described by Briggs and Williams (1981). Photography of the model from various angles was used to project the three-dimensional structure onto a two-dimensional plane—the photograph—mimicking the results of flattening in different orientations. The natural process of collapse and flattening differs only in that the elements would tend to rotate through the minimum angle necessary to bring their planar surfaces parallel to the bedding plane. Thus the foreshortening seen in the photographic images will not be apparent in bedding plane assemblages, but the relative positions of the elements should be replicated.

As our model is based on the standard pattern, direct lateral flattening, nearly perpendicular to the plane of symmetry, naturally replicates assemblages of that type. Variations on the pattern are produced by flattening from slightly oblique angles, offset upwards, forwards or backwards from the perpendicular, implying a tilting of the head at an angle to the bedding. In particular, the 'offset standard' pattern can be explained by oblique flattening with the apparatus tilted to one side and downwards (or upwards) anteriorly (Fig. 4.8). The parallel pattern is replicated when the model is photographed along its axis from an

antero-lateral (or postero-lateral) position (Fig. 4.9); not only does an alignment of the elements result, but they become crowded as in the assemblages. The perpendicular pattern is more difficult to replicate, but may be explained by a collapse of the apparatus that causes the pectiniform elements to fall in a different direction from the ramiform elements. This could be achieved with the head tilted forwards so that the pectiniform elements rotate parallel to the apparatus axis, and slightly to one side, causing the ramiform set to topple sideways (Fig. 4.10). This would occur only within a limited range of possible orientations, but perpendicular assemblages are known to be rare.

A crucial test of our model lies in its ability to explain the two seemingly incompatible linear patterns. The 'linear 1' arrangement, with the elements in their correct conventional orientations, is readily produced by dorso-ventral flattening, the head tilted slightly downwards anteriorly (Fig. 4.11). The 'linear 2' arrangement (where an anterior positioning of the ramiform set produces a reversed orientation of the elements) results when the head is tilted upwards anteriorly (Fig. 4.12). Flattening in this orientation leads to overlap of the Pa and Pb pairs, as seen in Du Bois' assemblage. The disruption of the ramiform

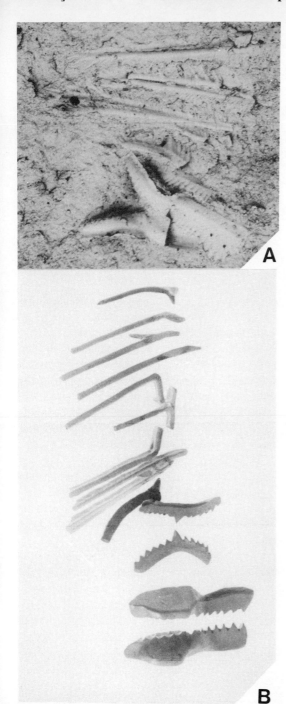

group in this assemblage suggests that these elements may have been oriented almost normal to the bedding plane. A reduced angle of upward tilting would result in a linear arrangement in which the pectiniform elements are oriented in the opposite direction from the ramiform elements. An assemblage that approaches this arrangement was illustrated by Schmidt (1934, fig. 3), but a component of lateral tilting has produced a pattern that is transitional between linear and perpendicular.

Flattening of our model thus accounts for all the repeated arrangements shown by bedding plane assemblages. The standard pattern, in particular, is shown to be produced by lateral compaction of a bilaterally symmetrical apparatus, without any 'rotation' or 'contraction'. It follows that the head, at least, of the first Granton animal is preserved in lateral aspect, and the apparent bilateral symmetry of the soft parts in that specimen is not indicative of dorso-ventral compaction (*contra* Bengtson 1983). It is probable that the entire specimen is in lateral compaction, confirming suggestions that the structures of the trunk are comparable with those of chordates (see also Aldridge and Briggs 1986, Aldridge *et al*. 1986). The predominance of standard patterns among known bedding plane assemblages indicates that, on death, polygnathacean conodont animals usually came to rest on the sediment lying on their sides. This implies that the animals were laterally flattened in life.

## 4.7   THE FUNCTION OF THE POLYGNATHACEAN APPARATUS

The architecture of polygnathacean apparatuses deduced herein and represented by our model is radically different from most reconstructions proposed hitherto. Previous authors have envisaged the elements as longitudinally aligned, often in a linear series (see Aldridge 1987, fig. 1.5 this volume). The evidence of bedding plane assemblages, on the other hand, indicates that the pectiniform elements were

Fig. 4.9—The parallel pattern: A, bedding plane assemblage from Bailey Falls, La Salle County, Illinois, specimen no. 57P–180I, Illinois State Geological Survey, ×25; B, postero-lateral view of model.

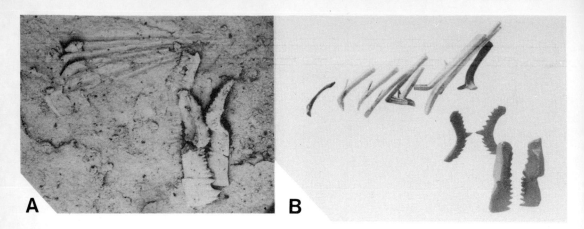

Fig. 4.10—The perpendicular pattern: A, specimen no. 57P–170II, ×23; B, oblique posterior view of model.

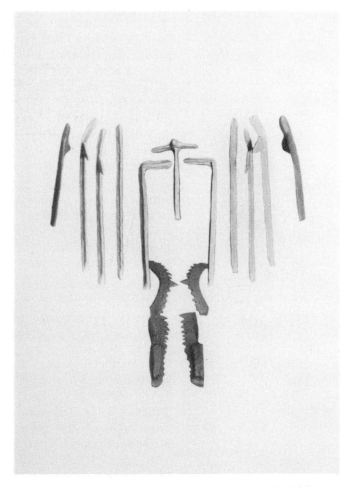

Fig. 4.11—The linear 1 pattern: model viewed from above and slightly to the anterior.

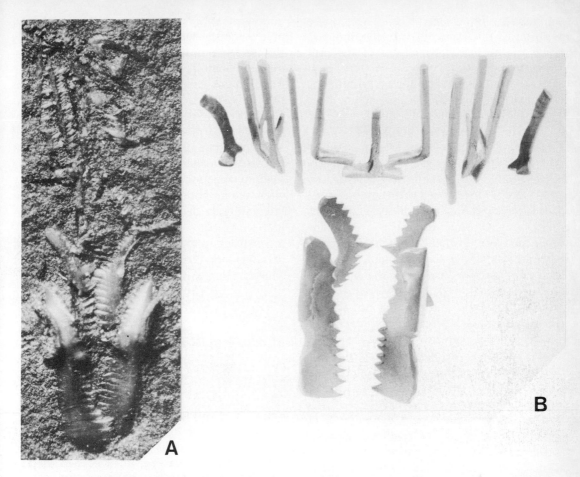

Fig. 4.12—The linear 2 pattern: A, specimen no. X-1480, ×25; B, model viewed from above and somewhat to the posterior.

transverse and the ramiform elements oblique to the axis of the animal (see also Norby 1976, p. 54). With the ramiform elements at the anterior, what are the implications of this arrangement for the function of the apparatus?

Briggs *et al.* (1983) considered that the most credible integrated functional system involved the ramiform elements in grasping prey which was processed by the pectiniform elements. It is possible to envisage the opposed denticulate surfaces of the pairs cutting and grinding, but the cusps and denticulated processes of the ramiform elements are not preserved in opposition, and could not have functioned in grasping in this attitude (Nicoll 1985). In order to

grasp they must have adopted a different position from that in which they are found in bedding plane assemblages and clusters. This idea has been forwarded several times; for example, Jeppsson (1971, p. 107) stated that 'it may be difficult to define one exact position of each element, but it will perhaps appear that they had different resting-growing positions and working positions'. To bring the cusps of the ramiform elements in our model into opposition would require a rotation of each side of this part of the apparatus inwards through about 90°. This action could have been accompanied by extrusion of the cusps from the enclosing epithelium in the manner postulated by Bengtson (1976). The mode of operation of this system is broadly analogous to that of the lingual apparatus of the myxinoids, in that both involve the bilateral operation of paired elements and incorporate functional and resting positions (Dawson 1963, Yalden 1985). It is likely the movement also brought the long axes of the elements into a longitudinal position to allow food to pass back to the pectiniform elements. We see no reason to propose that the pectiniform elements were rotated into a longitudinal orientation to function. However, elements of this type are rarely planar, but show some curvature of the blade, so that the entire element could not have been perpendicular to the food channel. The conventionally anterior portion of the units may have been normal to the body axis, causing the posterior blade or platform to be directed slightly backwards. Alternatively, each pectiniform element may have been slightly more inclined so that the posterior portion lay behind the anterior portion. This could explain the differentiation of Pa elements into anterior free blade and posterior platform.

There is no unequivocal evidence as to whether the cusps of the ramiform elements were directed dorsally or ventrally, but an indication is provided by the Granton animals. These show an asymmetrical development of fin rays on either side of the posterior extremity of the trunk (Aldridge *et al*. 1986, fig. 4). Only those of one side, perhaps that with the greater development, are evident on the first specimen (Briggs *et al*. 1983, figs. 4,5). This margin may represent the dorsal. There is no evidence of major twisting of the trunk, and if we trace this margin to the head of the animal (Fig. 1.9) the cusps of the ramiform elements and the platforms of the Pa elements would appear to be directed ventrally.

## 4.8 POSSIBILITIES FOR FURTHER RESEARCH

We have established an architecture and proposed a functional model for polygnathacean conodont apparatuses, consistent with available evidence. Details of the model, especially the relative positions of the ramiform elements, remain to be refined. In particular, the M elements commonly display an independence from the S elements, suggesting that they may have been operated by different muscles. The model can also be extended and tested by examining apparatuses with different elemental morphologies and relationships. The field of functional morphology is now truly open to the conodont worker, and it will be interesting to see to what extent the polygnathacean plan can serve as a template in reconstructing the architecture and operation of other non-coniform apparatuses, many of which include strikingly different elements.

## ACKNOWLEDGEMENTS

Our work on conodont palaeobiology is funded by N.E.R.C. Research Grant GR/3/5105. We thank Dr D. B. Blake, Department of Geology, University of Illinois, for arranging the loan of material from the Scott and Rhodes assemblage collections, and Dr L. Kent, Illinois State Geological Survey, for access to the Du Bois, Avcin, and Norby collections. Line drawings were drafted by Mrs J. Wilkinson, and photographic assistance was provided by Mr D. Jones and Mr A. Swift.

## REFERENCES

Aldridge, R. J. 1987. Conodont palaeobiology: a historical review. In: R. J. Aldridge (ed.): *Palaeobiology of Conodonts*, Ellis Horwood, Chichester, Sussex, 11–34.

Aldridge, R. J. and Briggs, D. E. G. 1986. Conodonts. In: A. Hoffman and M. H. Nitecki (eds): *Problematic fossil taxa*, Oxford University Press.

Aldridge, R. J., Briggs, D. E. G., Clarkson, E. N. K., and Smith, M. P. 1987. The affinities of conodonts—new evidence from the Carboniferous of Edinburgh, Scotland. *Lethaia* **19** 279–291.

Avcin, M. J. 1974. Des Moinesian conodont assemblages from the Illinois Basin. *Unpublished*

*PhD thesis*, University of Illinois at Urbana-Champaign, 152 pp., 5 pls.

Bengtson, S. 1976. The structure of some Middle Cambrian conodonts, and the early evolution of conodont structure and function. *Lethaia* **9** 185–206.

Bengtson, S. 1983. A functional model for the conodont apparatus. *Lethaia* **16** 38.

Briggs, D. E. G., Clarkson, E. N. K., and Aldridge, R. J. 1983. The conodont animal. *Lethaia* **16** 1–14.

Briggs, D. E. G. and Williams, S. H. 1981. The restoration of flattened fossils. *Lethaia* **14** 157–164.

Collinson, C., Avcin, M. J., Norby, R. D., and Merrill, G. K. 1972. Pennsylvanian conodont assemblages from La Salle County, northern Illinois. *Illinois State Geological Survey Guidebook Series* **10** 37 pp.

Dawson, J. A. 1963. The oral cavity, the 'jaws' and the horny teeth of *Myxine glutinosa*. In: A. Brodal and R. Fange (eds.): *The biology of* Myxine, Universitetsforlaget, Oslo, 231–255.

Du Bois, E. P. 1943. Evidence on the nature of conodonts. *Journal of Paleontology* **17** 155–159, pl. 25.

Hitchings, V. H. and Ramsay, A. T. S. 1978. Conodont assemblages: a new functional model. *Palaeogeography, Palaeoclimatology, Palaeoecology* **24** 137–149.

Jeppsson, L. 1971. Element arrangement in conodont apparatuses of *Hindeodella* type and in similar forms. *Lethaia* **4** 101–123.

Mashkova, T. V. 1972. *Ozarkodina steinhornensis* (Ziegler) apparatus, its conodonts and biozone. *Geologica et Palaeontologica* **SB1** 81–90, 2 pls.

Nicoll, R. S. 1985. Multielement composition of the conodont species *Polygnathus xylus xylus* Stauffer

1940 and *Ozarkodina brevis* (Bischoff and Ziegler 1957) from the Upper Devonian of the Canning Basin, Western Australia. *BMR Journal of Australian Geology and Geophysics* **9** 133–147.

Norby, R. D. 1976. Conodont apparatuses from Chesterian (Mississippian) strata of Montana and Illinois. *Unpublished PhD thesis*, University of Illinois at Urbana-Champaign, 303 pp., 21 pls.

Norby, R. D. 1979. Elemental architecture of natural platform conodont apparatuses of Mississippian and Pennsylvanian age. *Ninth International Congress of Carboniferous stratigraphy and geology, Urbana–Champaign, 1979, Abstracts of papers*, 249.

Rhodes, F. H. T. 1952. A classification of Pennsylvanian conodont assemblages. *Journal of Paleontology* **26** 886–901, pls. 126–129.

Robison, R. A. (ed.): 1981. *Treatise on Invertebrate Paleontology, Part W, Supplement 2, Conodonta*, Geological Society of America and University of Kansas Press, Lawrence, Kansas, 202 pp.

Schmidt, H. 1934. Conodonten-Funde in ursprünglichen Zusammenhang. *Paläontologische Zeitschrift* **16** 76–85, pl. 6.

Schmidt, H. and Müller, K. J. 1964. Weitere Funde von Conodonten-Gruppen aus dem oberen Karbon des Sauerlandes. *Paläontologische Zeitschrift* **38** 105–135.

Sweet, W. C. 1985. Conodonts: those fascinating little whatzits. *Journal of Paleontology* **59** 485–494.

Von Bitter, P. H. and Merrill, G. K. 1985. *Hindeodus, Diplognathodus* and *Ellisonia* revisited—an identity crisis in Permian conodonts. *Geologica et Palaeontologica* **19** 81–96, 1 pl.

Yalden, D. W. 1985. Feeding mechanisms as evidence for cyclostome monophyly. *Zoological Journal of the Linnean Society* **84** 291–300.

# 5

# Form and function of the Pa element in the conodont animal

R. S. Nicoll

## ABSTRACT

Blade and platform type Pa elements, both as fused pairs and in pairs reconstructed from discrete elements, have been examined in their presumed functional position with oral surfaces intermeshed. This has provided clues to the function performed by these elements in the conodont animal. Based mainly on Upper Devonian or Lower Carboniferous conodont species, four major categories of morphological relationship between Pa element pairs are recognized and defined: blades, plates, hollows, and crests. Cone and bar type Pa elements have not been studied.

Pa elements of the types studied are believed to have performed a crushing function on fine food particles passing through the mouth of the conodont animal. Variation of element morphology may be related to differences in the type of food particles consumed by different species of conodonts. The conodont animal was probably a microphagous feeder.

Amphioxus is suggested as a possible living relative of the conodonts. If they are related, amphioxus is presently restricted to a much narrower range of habitat than is envisioned for most conodont species.

## 5.1 INTRODUCTION

The relationship between conodont element morphology and function is an area of study that has been given relatively little emphasis compared with taxonomic, biostratigraphical, or ecological aspects. There has been a general tendency among those who study conodont elements to regard them as objects to be classified, and not to speculate about their possible function. In addition, most papers that have discussed the possible affinities of the conodont elements or animal have not carefully related element morphology to functional hypotheses.

There have been some exceptions, including the studies of Jeppsson (1971, 1979, 1980), Lindström (1974), Conway Morris (1976, 1980), Bengtson (1980, 1983a, 1983b), and Repetski and Szaniawski (1981). The recent description of an organism that really is the conodont animal (Briggs *et al*. 1983), after the descriptions of several animals that are most probably not (Melton and Scott 1973, Conway Morris 1976), has given us a body on, or in, which we may now with some conviction place the various elements that apparently formed the only mineralized structures of the animal.

However, the preservation of the first specimen of the animal is such that it leaves many questions unanswered, and if every conodont worker had a wish it would probably be to find a specimen that would give us an unsquashed view of the beast.

Even without the luxury of a perfect specimen it should be possible to hypothesize constructively on the probable function of the elements within the conodont animal. Jeppsson (1979) has suggested that the morphology of at least some elements implies a tooth function. Others, such as Conway Morris (1980) or Nicoll (1977, 1985) have rejected a tooth function. It should be possible to re-examine the evidence in the light of recent discoveries, and establish the function of the elements.

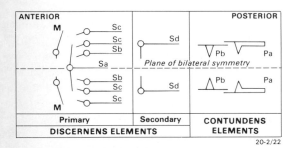

Fig. 5.1—The arrangement of discernens and contundens elements in the apparatus structure of the conodont animal.

The wide range of oral surface morphologies of conodont elements precludes a general statement about element function that will cover all element types of all genera. It is probable that the functions of discernens (ramiform) elements (Fig. 5.1) are quite different from those of the contundens (pectiniform) elements (Nicoll 1985). It is also not necessarily possible to homologize coniform element morphology and function with pectiniform–ramiform elements. For this reason coniform elements are not considered in the following discussion.

## 5.2  ELEMENT SEQUENCE

The order or sequence of elements in the conodont animal was a matter of dispute until the recovery of the specimen from the Granton shrimp bed of Edinburgh, Scotland (Briggs *et al*. 1983). Two sequences had been generally proposed, one with the Pa elements anterior (Hitchings & Ramsay 1978) and the other with the discernens elements anterior (Jeppsson 1971). The specimen described by Briggs *et al*. (1983) showed that the discernens elements are located at the anterior end of the apparatus and the contundens elements at the posterior (Fig. 5.1).

Bengtson (1983a) suggested that the apparatus in the specimen from the Granton shrimp bed is not in its functional position but in a resting position. However, the positions of the elements in this specimen are very similar to those observed in many of the fused clusters from the Silurian of northern Indiana (Pollock 1969, Nicoll and Rexroad 1987). This suggests that the position of the elements in the Granton specimen represents a normal distortion of the apparatus that may have resulted from sediment compaction or from factors related to the decomposition of the tissues of the organism.

Using Devonian fused clusters and the Granton specimen, Nicoll (1985) suggested that the order of elements in the apparatus was M elements at the anterior, followed by a complex of S elements, the Pb element pair, and lastly the Pa element pair (Fig. 5.1). This is essentially the same as the sequence proposed by Jeppsson (1971), and has been confirmed by the examination of additional cluster material (Nicoll and Rexroad 1987).

## 5.3  ELEMENT TYPES

There has been a gradual expansion of the number of elements included within the conodont apparatus, and it is now generally recognized that the maximum number of element types is seven. Some apparatuses may contain fewer element types, but it is doubtful that apparatuses will be found with more than seven, unless there is a differentiation of the two pairs of Sc elements. Thus many mid-Palaeozoic conodont apparatuses, exclusive of the Icriodontidae, consist of 15 elements representing seven element types as follows: M (2), Sa (1), Sc (4), Sb (2), Sd (2), Pb (2), and Pa (2).

The earliest conodonts, proto-, para-, or eu-, have simple undifferentiated cone morphologies. By the late Cambrian there was a differentiation of element types within the apparatus structure, and by the Early Ordovician it is possible to distinguish as many as seven distinctive element types in each apparatus. This number appears to hold for the rest of the time range of the Conodonta as the maximum found in any one animal.

The differentiation of element morphology within each individual animal probably represents the initiation of specialization of element function within the apparatus structure. Some element types appear to be very conservative; for example, it is possible to recognise an M element because there is a consistency about its morphology regardless of the species to which it is assigned. This conservatism of morphological type is a good indication that the function of the M element did not change significantly after the basic form evolved in the Early Ordovician.

## 5.4  THE Pa ELEMENT

If the M element represents a conservative element form, then the Pa element is the radical element. No other element types evolved anywhere near the number of morphological styles that can be found in this group.

The Pa element is usually considered to be the most important element in the conodont apparatus. It appears to be the element that evolved most rapidly, and has thus come to be the one on which most conodont biostratigraphy has been based. Morphologically the Pa element is usually more complex than the other elements of the apparatus, and is the most differentiated. Frequently the Pa element has developed lateral processes and/or expanded laterally to form a platform.

The Pb element is structurally linked with the Pa element. It shares a similar transverse orientation (Nicoll, 1985) and appears to have been located slightly anterior to the Pa elements, but may have been partly overlapping. The function of the Pb element is not the same as that of the Pa element, and it has not developed as great a range of morphologies. It is thus relatively conservative.

The reason for the preferential morphological malleability of the Pa element over the rest of the elements in the apparatus structure is not totally understood. I believe that it is probably a response to variation in the prey that the animal was feeding upon. There may be an analogy between conodont elements and teeth in those vertebrates that chew or grind their food, where the rear teeth are usually the most complex because they are assigned the task of breaking down the lumps that are passed back by the gathering and cutting front teeth. Rear teeth reflect closely the type of food eaten by the animal, and are modified from species to species as eating habits vary.

There is a broad evolutionary progression in the morphological change of the Pa element. It started in the Cambrian where, if it can be differentiated, the element had a coniform morphology. In the Ordovician the denticulated bar was introduced, and this was soon followed by the blade. The platform was the last general type to be introduced as the denticulated bars and blades developed lateral expansions. By the late Devonian or early Carboniferous the coniform and denticulated bar Pa element morphologies were rare, and the blade and platform morphologies were dominant till the late Triassic. The discussion that follows deals only with the blades and platforms.

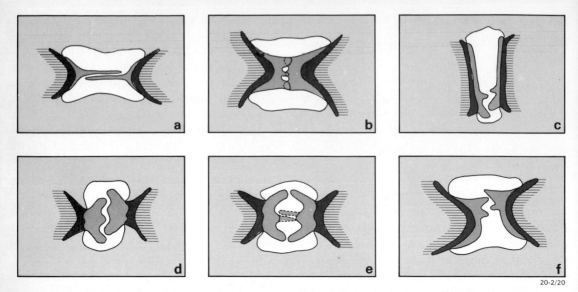

20-2/20

Fig. 5.2—Cross-sections of the major categories of Pa element described in the text. Each example shows the element crown in the mouth opening with the basal plate (solid black) and connective tissue (lines) embedded in body tissue. Tissue covering the crown is not shown. Shapes of the cross-sections are schematic but are based on Pa elements of genera as follows: a, blade form – *Ozarkodina*; b, plate form – *Icriodus*; c, plate form – *Palmatolepis*; d, plate form – *Polygnathus*; e, hollow form – *Polygnathus*; f, crest form – *Gnathodus*.

Fused clusters of Silurian blade-type *Ozarkodina* Pa elements (Pollock 1969, Nicoll and Rexroad 1987) show that the most probable life orientation of these elements was with their inner lateral faces overlapping (Fig. 5.2a). Assuming that the elements functioned in place within the animal, and were not everted, the elements could have moved apart and together with a cutting action, or have had a crushing motion with the inner lateral faces in near contact. I consider that the crushing motion was the effective action of the elements. This action could have mashed the food particles by rolling them between the element faces, thus crushing the outer cell walls or tests of the plankton that were the probable prey of the conodont animal. It is important to remember that the elements were most probably tissue-covered, and thus the hard surface of each element was not in contact with the opposing element.

The platform-type Pa elements have much more complex inter-element working relationships. In general the working surface becomes the face of the platform, but in detail the relationship is much more complex. Three broad categories of platform upper surfaces can be recognized: the plate, the hollow, and the crest (Fig. 5.2). The plate form has a large area of surface contact, and the element surfaces can be either smooth or highly irregular. The hollow form has an open area along the midline of the elements with the lateral margins in near contact. The crest form is the opposite of the hollow form with only a small or narrow area along the midline of the elements in contact and the lateral margins of each element curving away from the opposing element. These are arbitrarily defined categories, and intergradational forms exist between all of the end-points.

The changes in element morphology between these three types are not great, but they are different enough to indicate some variation in the way food particles were processed by different groups of conodont animals.

## 5.5 EXAMPLES

I have selected several Pa elements drawn from five genera, *Polygnathus, Palmatolepis, Gnathodus, Ozarkodina*, and *Icriodus*, to illustrate some of the relationships in the different morphological categories (see Fig. 5.2).

**(a)** *Ozarkodina brevis* **(Bischoff and Ziegler 1957) and** *O. eosteinhornensis* **(Walliser 1964).**

The simple blade morphology of *Ozarkodina* Pa elements means that there is no oral surface to mesh with the opposing element, and, unless the elements functioned as cutting blades, it would have been the lateral faces of the overlapping elements which were the functional surfaces. Two examples, both fused clusters, are shown (Pl. 5.1, Figs. 1, 2). Despite their different morphologies (*O. brevis* has a tall cusp) both show the typical blade tip to cavity flare overlap of this category. There does not appear to be any difference between the inner and outer surfaces of the blade elements, that is between those that were in contact with the opposing element and those that were not.

The *Ozarkodina* blade-type morphology must have been a very successful element style because it remained basically unchanged in a direct evolutionary succession from the Late Ordovician to the Carboniferous. Yet *Ozarkodina* was also the stock from which many, or most, of the Silurian to Carboniferous platform style genera were directly or indirectly derived.

**(b)** *Icriodus expansus* **Branson and Mehl 1938**

The Icriodontidae have a very distinctive oral surface morphology, with some genera having a transverse row of three denticles, added in growth as a triad (Nicoll 1983). By using some of the numerous cluster pairs of I elements obtained from the Canning Basin (Nicoll 1983), it is possible in this case to be sure that the two I elements really did fit together in

life (Pl. 5.1, Figs. 3, 4). The oral surfaces show that when one element has a distinctive denticle triad, for example with one or both of the outer denticles enlarged, the opposing element has an equivalent enlarged gap between the denticles of the triads, allowing the elements to intermesh. This intermeshing means that particulate matter passing across the element surface could be very finely mashed, almost chopped. It is interesting to note that the denticles of juvenile *Icriodus* I elements are usually pointed, but those of adult elements are normally rounded. If the function of the elements was to cut, one might expect the sharp form to have been retained.

**(c)** *Palmatolepsis tenuipunctata* **Sannemann, 1955**

The large Pa elements of *Palmatolepis* are good examples of plate morphology, with a large surface area, which is generally smooth. The carina of these elements stands far enough above the plate so that the greater part of the plate surfaces cannot have been in direct contact with the opposing element (Pl. 5.2, Figs. 1–4).

**(d,e)** *Polygnathus xylus xylus* **Stauffer 1940;** *P. webbi* **Stauffer 1938;** *P. parapetus* **Druce 1969;** *P. siphonellus* **Druce 1969**

The specimens of *Polygnathus* chosen are assigned to either the plate or hollow morphological groups. *P. xylus xylus* and *P. webbi* show plate morphology (Pl. 5.1, Figs. 5–8), and *P. parapetus* and *P. siphonellus* have a hollow morphology (Pl. 5.3, Figs. 3–12). The plate morphologies show very little open space between trough and ridge when the opposing elements are closely appressed. The hollow forms have a large open area in the central part of the structure, and the carina is usually very low. Rib ornamentation on the platform oral surface may be either transverse or longitudinal.

**(f) *Gnathodus bilineatus modocensis* Rexroad 1957**

Representatives of the crest morphology are not very common. The specimen of *G. bilineatus modocensis* illustrated shows the typical high carina ridge and the shoulder platform over the basal cavity (Pl. 5.3, Figs. 1, 2). The surfaces of this shoulder platform cannot be in contact with the opposing element surface, and, unlike elements of the hollow morphology, there is no confining opposite lip or surface. Unless confined by soft tissue lining the mouth cavity it would seem that most of the food particles could easily escape any effective action performed by elements of this type.

## 5.6 Pa ELEMENT FUNCTION: CONSTRAINTS

If the basic function of Pa elements of different species is the same, or at least similar, then there was a wide range of morphological solutions for implementation of that function. I have illustrated some of the more common types of middle Palaeozoic element morphologies found in platform-type Pa elements,

but there are other platform and numerous non-platform Pa elements that represent a much wider range of morphological styles.

It is important to state here that I consider the conodont animal to have been most probably microphagous rather than macrophagous. This distinction is important in the interpretation of how the elements functioned. In a macrophagous organism the teeth or tooth-like structures are usually associated with prey capture, as well as the mechanics of size reduction of the prey in the ingestion process. In microphagous organisms the teeth or tooth-like structures are usually associated with prey selection or rejection and may be essentially passive structures with relatively little movement. They are effectively sieves. Food particle movement in microphagous organisms is usually controlled by water currents established by cilia in various parts of the mouth or pharynx.

If it is assumed that the basic function of the Pa element was to assist in the ingestion of food, then the variation in element morphology must be related either to differences in the type of food consumed or to the method by which the Pa element assisted in the processing. Because most of the examples considered

Plate 5.1—All ×65.
Specimens are resposited either in the Commonwealth Palaeontological Collection (CPC), Bureau of Mineral Resources, Canberra, Australia or the Indiana University – Indiana Geological Survey Collection (IU-IGS), Bloomington, Indiana, USA.
Fig. 1—*Ozarkodina brevis* (Bischoff and Ziegler 1957), stereo photomicrographs of fused Pa element pair, blade form, lateral view, CPC25202, WCB 804/5, Canning Basin, Western Australia; Napier Formation, Upper Devonian.
Fig. 2—*Ozarkodina eosteinhornensis* (Walliser 1964), stereo photomicrographs of fused Pa element pair, blade form, lateral view, IU–IGS16829, France Stone Company quarry, Cass County, Indiana; Salina Formation, upper Silurian.
Figs 3, 4—*Icriodus expansus* Branson and Mehl 1938, stereo photomicrographs of fused I element pair, plate form, with adhering cone elements, oral/aboral views. Enlarged nodes of one element fit into internode space of opposing element. CPC25719, WCB 804/5, Canning Basin, Western Australia; Napier Formation, Upper Devonian.
Figs 5, 6—*Polygnathus xylus xylus* Stauffer 1940, fused Pa element pair, plate form. 5, stereo photomicrographs, lateral view. 6, posterior view. In life the right element would have been directly opposed and more closely fitted into left element. CPC25176, WCB 804/5, Canning Basin, Western Australia; Napier Formation, Upper Devonian.
Figs 7, 8—*Polygnathus webbi* Stauffer 1938, reconstructed element pair, plate form, CPC25720 (left) and CPC25721 (right). 7, stereo photomicrographs, lateral view. 8, posterior view. Note the amount of separation of the posterior tips of the elements and the offset of the line of the carinas. Platform margin of right element slightly overlaps margin of left element. WCB308/27, Canning Basin, Western Australia; Napier Formation, Upper Devonian.

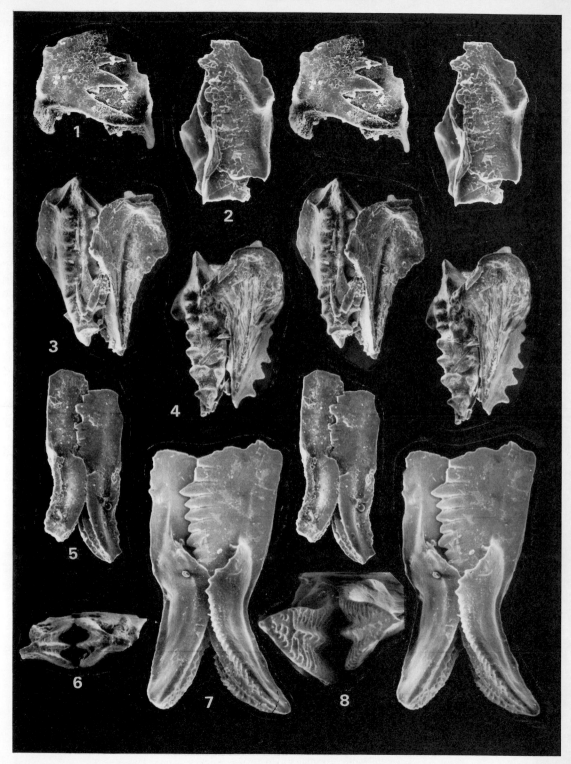

here are of closely related genera or species, it may be differences in food type that explain the differences in morphology.

It has been pointed out elsewhere (Nicoll 1985) that the morphological differences between the discernens elements and the contundens elements argues for a major difference in function between these two groups. Only the contundens elements, and usually only the Pa element, show variations in element morphology that indicate that the elements functioned in opposition.

Bengston (1983a) and others have suggested that when the conodonts elements were functioning, they were everted and used for prey capture. Only when the elements were in a resting position were they returned to their protective skin folds where growth and repair took place. This type of usage would probably not produce a high degree of morphological interrelationship of opposing elements, because there is no need for the elements to interdigitate when in action, nor would they need to be in close proximity in the resting state. However, if the morphology of the interrelating elements is indicative of their functioning in opposition, the Pa elements must have effectively worked toward each other, in the manner suggested in Fig. 5.3a.

There are other constraints on functional interpretation. These are that the growth accretion of the elements is on the outer surface, that the elements must have functioned during the growth period, and that there is a lack of evidence of wear on the outer surface. To account for all of these factors it has been suggested that the Pa elements were covered by tissue at the time of their use (Lindström 1974, Nicoll 1977). Conway Morris (1980) has pointed out some of the difficulties of trying to

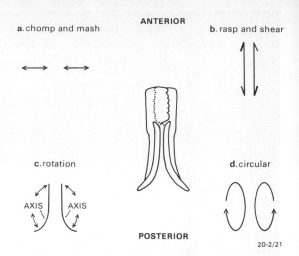

Fig. 5.3—Examples of the possible action of Pa element pairs of the general types discussed in the text. All movement is restricted to the plane of the paper by the ridge morphology of the carina which effectively prevents lateral motion of the elements.

return elements to a skin-fold pocket that contained phosphate-secreting cells.

## 5.7  Pa ELEMENT FUNCTION: INTERPRETATION

The Pa elements are the posterior elements of the mineralized apparatus of the conodont animal. This structure was located near the anterior end in the head region of the animal (Briggs *et al.* 1983). The anterior part of the apparatus, consisting of the discernens elements, functioned as a sieve to sort prospective food particles from non-food particles obtained from the water through which the animal was swimming (Nicoll 1985). The discernens elements were oriented with their oral surfaces directed ventrally, and did not function with oral surfaces intermeshed. Ciliated tissue

---

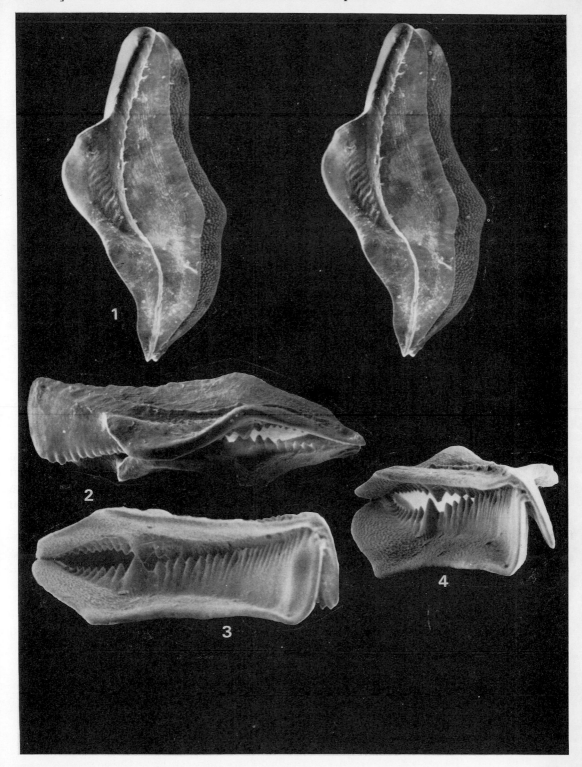

covering these elements moved food particles posteriorly towards the opening of the gullet.

The contundens elements are located just behind the discernens elements (Briggs *et al*. 1983, Nicoll and Rexroad 1987). The Pb elements are anterior of the Pa elements, but there may be some overlapping of the anterior end of the blade of the Pa element with the posterior end of the Pb element. Both Pb and Pa element pairs were oriented in a transverse fashion with the oral surface of each element pointed towards the opposing element. Clusters seem to indicate that elements in both the Pb and Pa pairs were partly overlapped by the opposing element (Nicoll and Rexroad 1987).

The width of the conodont animal is given as about 1.8 mm for the first specimen recovered from Granton (Briggs *et al*. 1983). The figure of 1.8 mm will represent a maximum dimension of the animal and may include some distortion caused by post-mortem compression. The width will have varied between species and with specimen age, but 1.8 mm will be used in the following calculations.

The Pa elements of the *Polygnathus* species used in this study have an average blade overlap of about 0.25 mm when the elements are in their closest possible orientation. Thus for the upper surface of the blade to have acted as a cutting tool the elements would have had to have moved apart in excess of 0.25 mm. When the Pa elements were appressed they took up about 0.5 mm of the 1.8 mm width of the

animal. With the Pa elements separated to a cutting position the width of the animal would have been at least 0.3 mm greater, or about 2.1 mm. It is unlikely that this degree of expansion could have been accommodated within the resting body width of the animal. Four layers of integument (skin) and the muscle and connective tissue associated with movement of the animal and the elements would easily fill the 1.3 mm of space surrounding the mouth–gullet opening. The body of the animal must thus have been flexible enough for some expansion, so that the contundens elements could have moved apart. If the elements are not called on to have a cutting function, it is still necessary for them to move apart sufficiently to allow passage of the food particles, but only enough to account for the diameter of the food particle which was probably not more than 0.1 mm. This particle size is determined from the spacing of the discrete denticles of the discernens elements, between which the particles would have to pass.

Jeppsson (1979) concluded that conodont elements had a tooth function, based on the similar morphologies of conodont elements and some types of vertebrate teeth. Other studies (Nicoll 1977, 1985) have suggested that they did not have a tooth function. The morphology of the Pb, and to some extent the Pa, elements does suggest a cutting function. However, the relationship of Pb elements in fused clusters suggests that the elements

---

Plate 5.3—All ×65.

Figs 1, 2—*Gnathodus bilineatus modocensis* Rexroad 1957, reconstructed element pair, crest form, CPC25724 (left) and CPC25725 (right). 1, stereo photomicrographs, later view. 2, posterior view. Note separation of shoulders of flared basal cavities that cannot have been in contact when elements were functioning. Abandoned quarry north of Pella, Marion County, Iowa; Pella Formation, Mississippian.

Figs 3–7—*Polygnathus parapetus* Druce 1969, reconstructed element pair, hollow form, CPC25726 (left) and CPC25727 (right). 3, anterior view of element pair. 4, posterior view of element pair. 5, lateral view of left element. 6, lateral view of element pair. 7, lateral view of right element. Note that the left margin of each element is significantly modified to allow the enlarged right margin to overlap when the elements are placed in their functional position. BGB 608/42, Bonaparte Basin, Western Australia; Burt Range Formation, Lower Carboniferous.

Figs 8–12—*Polygnathus siphonellus* Druce 1969, reconstructed element pair, hollow form, CPC25728 (left) and CPC25729 (right). 8, posterior view of element pair. 9, oral view of left element. 10, outer lateral view of element pair. 11, oral view of right element. 12, anterior view of element pair. Note amount of separation of carinas and groove on outer margin of left element that allows right element to mesh tightly. BGB 607/152, Bonaparte Basin, Western Australia; Burt Range Formation, Lower Carboniferous.

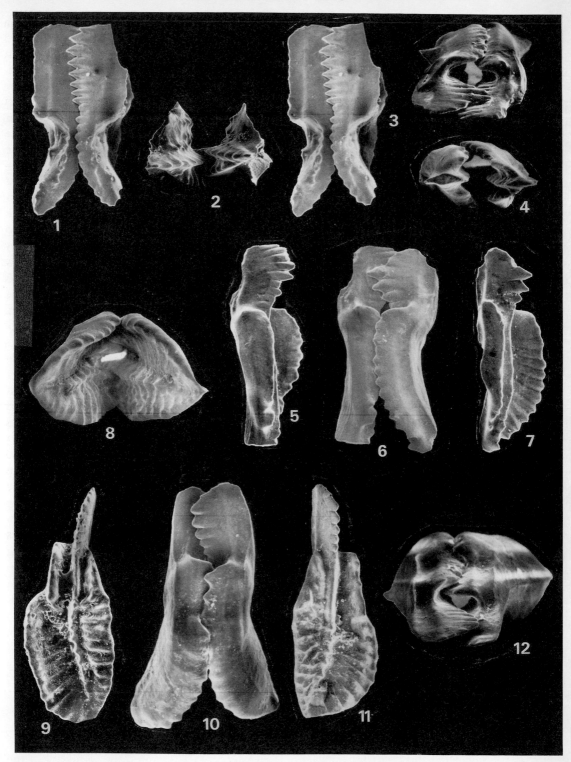

slipped past each other and the potential cutting surfaces of the upper margin could not effectively cut. Also, given the size restrictions on potential prey, probably not more than 0.1 to 0.2 mm, it is questionable if the food would need to be cut.

If the motion of both Pb and Pa elements is similar to that suggested in Fig. 5.3a, where the elements pull away from each other and then move together, each slipping laterally past the opposing element, the effective action of the elements is going to be that of rolling any particulate matter caught between them. This would bruise or crush the particulate matter, but would not cut it. The development of the platform of the Pa element is a result of increasing the efficiency of the crushing action. Particles would be pressed by the opposing platform surfaces and, again, crushed but not cut.

## 5.8   CONCLUSION

I think it is more important to construct an internally consistent model of the conodont animal than to force our interpretation into some existing group of organisms just because there is a superficial resemblance. It is important to keep our options open and not, at this stage, to commit our assignment of the conodont animal too closely to any one group. The recovery of additional specimens from the Granton shrimp bed locality holds the best bet for a resolution of the problem.

In recent years, chaetognaths and amphioxus have been frequently mentioned as possible living relatives of conodonts. The feeding habits of these two groups of living organisms are very different, as chaetognaths are macrophagous and amphioxus are microphagous. The function interpreted for conodont elements will be dependent on the group to which the conodonts are believed to be most closely related.

Szaniawski (1982), Bengtson (1983a), and others have suggested homologies between conodonts and chaetognaths, based mainly on the undoubted morphological similarity of chaetognath spines to protoconodont elements recovered from samples of Cambrian age. They have not adequately discussed some of the major problems of the chaetognath model. One of these problems is that the composition and mode of growth of the chaetognath spines are incompatible with those of conodont elements. Another is the difficulty of returning the conodont element to the skin-fold pocket were growth took place after its postulated exposure at the time of use.

Briggs *et al*. (1983) discussed a number of possible affinities of the conodonts based on preliminary interpretation of the soft-tissue morphology of a single specimen recovered from the Granton shrimp bed. They did not suggest a definite relationship with any group, but did opt for a tooth function of the elements.

The presence or absence of permanent tissue cover over the phosphatic structure of the element would greatly affect the interpretation of element function. If the element functioned without tissue cover a grasping function would seem probable. If elements were tissue-covered a filtering function would be more likely. Briggs *et al*. (1983) tended to equate tissue cover with the suggestion by Conway Morris (1976) of a tentacle-support structure rather than the proposal by Nicoll (1977) that the element may have served as support for ciliated tissue. Bengtson (1980), in discounting tissue cover, overlooked a number of organisms, such as the crinoids, that have a ciliated food-gathering system developed on a mineralized surface which captures food particles from the water and transports them into the mouth. Remember, too, that the turtle lacks teeth in the generally accepted sense, but has a very effective cutting structure that consists of chitinous tissue overlying bone.

The Cephalochordata (e.g. amphioxus) is another group of organisms that offers some potential as a model for conodont affinities. Briggs *et al*. (1983) noted the similarity of the apparent segmentation of the trunk of their specimen to the myotomes of amphioxus. Other areas of similarity are the buccal cirri and the wheel organ. The buccal cirri act as an initial sieve regulating the entry of particulate matter, and the wheel organ is a ciliated structure that catches and directs a mixture of food particles and mucus into the mouth of amphioxus (Barrington 1965). These structures lack mineralization but could be analogues to the discernens and contundens elements respectively. If conodonts were related to the Cephalochordata, they represent an advanced form; and the present example, amphioxus, is degenerative, that is more restricted in physiology, morphology, and ecology.

For several reasons amphioxus would be an attractive candidate for a modern relative of the conodont. The larval stage of some species of amphioxus may retain a pelagic form for as long as 200 days and move over distances of 8000 km (Webb 1975). This could help explain the world-wide synchroneity of conodont ranges. The life span of some amphioxus may be as much as 8 years (Courtney 1975), and with continuous growth this could account for the size range found in some conodont faunas.

In summary, the following statements may serve as guidelines for further discussion of Pa element function:

(a) The elements were located in near proximity to each other, at least at the time of the death of the organism, as proved by the Granton shrimp bed conodont animal (Briggs *et al*. 1983) and fused clusters (Nicoll and Rexroad 1986).

(b) Differences in element morphology suggest that the function of the discernens and contundens elements were different.

(c) The different styles of Pa element discussed in this chapter had a generally similar function.

(d) That function involved the ingestion of food and probably the crushing of prey. Other styles of Pa element, such as cones and denticulate bars, would have performed a slightly different function.

(e) Variations in the morphology of Pa elements are probably a reflection of the differing characteristics of preferred prey species.

(f) Food for the conodont animal was probably microplankton, and the animal was an active predator. A microphagous feeding strategy helps to explain the use to which the elements were adapted.

(g) Amphioxus is a possible modern relative of the conodont animal. If there is a relationship between amphioxus and the conodonts, it is clear that the present life style of amphioxus is much more constrained than that to which conodonts of Cambrian to Triassic age aspired.

## ACKNOWLEDGEMENTS

Peter J. Jones (Bureau of Mineral Resources, Canberra) critically read the draft manuscript and made numerous helpful suggestions. SEM photography and plate preparation were completed with the invaluable assistance of Arthur T. Wilson (B.M.R.). Publication was approved by the Director, Australian Bureau of Mineral Resources, Geology and Geophysics, Canberra.

## REFERENCES

Barrington, E. J. W. 1965. *The biology of Hemichordata and Protochordata*. Oliver and Boyd, Edinburgh, 176 pp.

Bengtson, S. 1980. Conodonts: the need for a functional model. *Lethaia* **13** 320.

Bengtson, S. 1983a. A functional model for the conodont apparatus. *Lethaia* **16** 38.

Bengtson, S. 1983b. The early history of the Conodonta. *Fossils and Strata* **15** 5–19.

Briggs, D. E. G., Clarkson, E. N. K., and Aldridge, R. J. 1983. The conodont animal. *Lethaia* **16** 1–14.

Conway Morris, S. 1976. A new Cambrian lophophorate from the Burgess Shale of British Columbia. *Palaeontology* **19** 199–222.

Conway Morris, S. 1980. Conodont function: fallacies of the tooth model. *Lethaia* **13** 107–108.

Courtney, W. A. M. 1975. The temperature relationships and age-structure of North Sea and Mediterranean populations of *Branchiostoma lanceolatum*. In: E. J. W. Barrington and R. P. S. Jefferies (eds.): *Protochordates. Symposia of the Zoological Society of London* **37** 213–233.

Hitchings, V. J. and Ramsay, A. T. S. 1978. Conodont assemblages: a new functional model. *Palaeogeography, Palaeoclimatology, Palaeoecology* **24** 137–149.

Jeppsson, L. 1971. Element arrangement in conodont apparatuses of *Hindeodella* type and in similar forms. *Lethaia* **4** 101–123.

Jeppsson, L. 1979. Conodont element function. *Lethaia* **12** 153–171.

Jeppsson, L. 1980. Function of the conodont elements. *Lethaia* **13** 228.

Lindström, M. 1974. The conodont apparatus as a food-gathering mechanism. *Palaeontology* **17** 729–744.

Melton, W. G., and Scott, H. W. 1973. Conodont-bearing animals from the Bear Gulch Limestone, Montana. In F. H. T. Rhodes (ed.): *Conodont Paleozoology. Geological Society of America Special Paper* **141** 31–65.

Nicoll, R. S. 1977. Conodont apparatuses in an Upper Devonian palaeoniscoid fish from the Canning Basin, Western Australia. *BMR Journal of Australian Geology and Geophysics* **2** 217–228.

Nicoll, R. S. 1983. Multielement composition of the conodont *Icriodus expansus* Branson and Mehl from the Upper Devonian of the Canning Basin, Western Australia. *BMR Journal of Australian Geology and Geophysics* **7** 197–213.

Nicoll, R. S. 1985. Multielement composition of the conodont species *Polygnathus xylus xylus* Stauffer, 1940 and *Ozarkodina brevis* (Bischoff and Ziegler, 1957) from the Upper Devonian of the Canning Basin, Western Australia. *BMR Journal of Australian Geology and Geophyscis* **9** 133–147.

Nicoll, R. S. and Rexroad, C. B. 1987. Re-examination of Silurian conodont clusters from northern Indiana. In: R. J. Aldridge (ed.): *Palaeobiology of Conodonts*. Ellis Horwood, Chichester, Sussex, 49–61.

Pollock, C. A. 1969. Fused Silurian conodont clusters from Indiana. *Journal of Paleontology* **43** 929–935.

Repetski, J. E. and Szaniawski, H. 1981. Paleobiologic interpretation of Cambrian and earliest Ordovician conodont natural assemblages. In: M. E. Taylor (ed.): *Short Papers for the Second International Symposium on the Cambrian System, 1981. United States Geological Survey Open-File Report* 81–743, 169–172.

Szaniawski, H. 1982. Chaetognath grasping spines recognised among Cambrian protoconodonts. *Journal of Paleontology* **56** 806–810.

Webb, J. B. 1975. The distribution of amphioxus. In: E. J. W. Barrington and R. P. S. Jefferies (eds.): *Protochordates. Symposia of the Zoological Society of London*, **36**, 179–212.

# 6

# A conodont animal from the lower Silurian of Wisconsin, USA, and the apparatus architecture of panderodontid conodonts

M. P. Smith, D. E. G. Briggs, and R. J. Aldridge

## ABSTRACT

A conodont animal of the genus *Panderodus* from the Brandon Bridge of Waukesha, Wisconsin, displays the anterior portion of a segmented trunk and an assemblage in which the elements are preserved in opposed pairs. The configuration of the elements, which are in linear sequence with the cusps pointing adaxially and posteriorly, provides several constraints on the architecture of the *Panderodus* apparatus. Although the single specimen cannot provide unequivocal evidence, the assemblage suggests that the apparatus comprised two opposed halves, operating bilaterally with the elements arranged in planar or arched linear arrays. This architecture is consistent with evidence from fused clusters, which also indicate that the apparatus comprised seven pairs of elements. It is possible that each half-apparatus was supported by a contiguous basal structure.

## 6.1 INTRODUCTION

The majority of conodont natural assemblages have been recovered from Upper Palaeozoic strata, so most attempts at reconstructing apparatus architecture have concentrated on taxa of that age. The evidence from older rocks is, in contrast, scanty, and until Mikulic *et al.* (1985a,b) reported a *Panderodus* assemblage from the Llandovery of Waukesha, Wisconsin, not a single Lower Palaeozoic euconodont bedding plane assemblage was known. However, fused clusters of coniform elements are occasionally encountered, and representatives of several genera have been described, including *Belodina* (Barnes 1967, Nowlan 1979), *Besselodus* (Aldridge 1982), *Cordylodus* (Repetski 1980), *Drepanoistodus* (Smith 1985), *Parapanderodus* (Smith 1985), *Proconodontus* (Repetski 1980), and *Scolopodus* (An *et al.* 1983), as well as *Panderodus*. Of these, clusters of *Panderodus* are the most common and have been described from many areas, including Indiana (Pollock 1969), Podolia (Dzik and Drygant 1986), and North China (An *et al.* 1983).

Unfortunately, clusters provide a less reliable basis for restoring skeletal architecture than bedding plane assemblages, as the apparatuses are rarely complete and it may also be difficult to assess the extent of any post-mortem disorganization. Two of the known clusters are, however, more complete than the majority. That of *Besselodus arcticus*

Aldridge 1982, from the Late Ordovician of North Greenland, is probably an entire half-apparatus, with a series of laterally fused elements including an oistodiform element at one end of the row. The *Panderodus* cluster described by Dzik and Drygant (1986) from the Llandovery of Podolia seems to have only one element missing, although some post-mortem disorganization of the assemblage is evident. The *Panderodus* bedding plane assemblage from the Brandon Bridge argillaceous dolomite of Waukesha (Mikulic *et al.* 1985a,b) is, however, of particular importance, as it provides the most reliable evidence to date for the architectural restoration of coniform apparatuses. Furthermore, the presence of soft parts, albeit poorly preserved, permits preliminary comparisons with the better-preserved polygnathacean animals from the Carboniferous of Granton, Scotland (Briggs *et al.* 1983, Aldridge *et al.* 1986).

## 6.2  SOFT PARTS OF THE WAUKESHA ANIMAL

Only the anterior part of the body is preserved on the edge of a slab; the posterior part was not found. The soft parts are preserved as a thin layer of white mineral similar to that analysed as fluorapatite in a leperditicopid ostracod

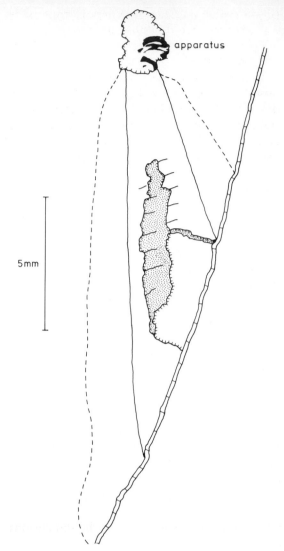

Fig. 6.2—Camera-lucida drawing of *Panderodus* conodont animal, UW4001/7a, anterior upwards; showing position of apparatus and segmentation of trunk.

from the same locality. The mineralized area does not extend as far as the assemblage (Figs. 6.1, 6.2), but a number of lines of evidence suggest that they are integral (Mikulic *et al.* 1985a):

(i)   the assemblage is positioned at the end of the body trace.

(ii)  the apparatus and preserved soft parts share the same axis of symmetry.

Fig. 6.1—*Panderodus* conodont animal from Waukesha County, Wisconsin, anterior to left; UW4001/7a, Geology Museum, University of Wisconsin, Madison, ×6. Specimen first illustrated by Mikulic *et al.* (1985a, fig. 2G).

(iii) a reduction halo, pale grey in contrast to the general reddish colour of the rock, envelopes both the soft parts and the assemblage.

Within the halo, a dark grey area of slightly raised relief extends from the fluorapatitic patch to the assemblage. At the edge of the slab, the white fluorapatite can be seen to extend beneath, but not beyond, the dark grey matrix. This area, therefore, probably represents the extent of the underlying fluorapatite and delimits the body trace. If this is so, then the soft parts are narrowest immediately behind the apparatus where the trunk is 1.3 mm across. As the animal is in dorso–ventral compaction (see section 6.6) this measurement represents the width of the fossil at this point. The trunk widens posteriorly and attains a maximum width of 3.4 mm before one margin is truncated by the edge of the slab. The body walls are straight; extrapolation of the shorter margin beyond the edge of the slab suggests that the width of the trunk (represented by only one margin) may have reached at least 5 mm. This contrasts with the trunk of the Granton animals, which are all preserved in lateral aspect (Aldridge *et al*. 1986). None of these specimens is higher than 2.2 mm, and, although they show some narrowing behind the head this never exceeds 50% of the total height. Although the Waukesha and Granton specimens are compacted in different orientations to bedding, the dimensions of the former suggest a larger and more tapered trunk.

Traces of segmentation in the Waukesha animal are evident in the exposed area of fluorapatite and in the immediately adjacent, dark grey, overlying matrix (Figs. 6.1, 6.2). At least seven somites are delimited, and extrapolation along the trunk suggests that there could have been more than 20 in the preserved section. The boundaries are straight and near normal to the sagittal axis, in contrast to the V-shaped segmentation of the Granton specimens (Aldridge *et al*. 1986).

## 6.3 DEVELOPMENT OF THE SPECIES CONCEPT IN *PANDERODUS*

Prior to 1966 the elements of *Panderodus* were described in form taxonomy, with distinctive morphological types given separate specific names. Bergström and Sweet (1966) produced the first multielement reconstruction when they recognized that two element types, broad-based forms referred to *P. compressus* and narrow-based costate forms referred to *P. gracilis*, belonged to a single species. This two-element concept was maintained by several subsequent workers (Cooper 1975, 1976, Barnes 1977, Barnes *et al*. 1979), who proposed different designations for the element types (Fig. 6.3). Cooper (1975, 1976) recognized that a more extensive suite of element types could be distinguished within the narrow-based costate category, but Barrick (1977) was the first to describe them. He attempted to demonstrate homologies between these elements and those of the symmetry transition series of ramiform apparatuses, applying the notational scheme devised by Sweet and Schönlaub (1975). He considered that the broad-based element was an M element, but gave no reason for rejecting the possibility that it represented a P element (cf. Cooper 1981).

Sweet (1979), while recognizing that the narrow-based, costate forms could be subdivided, did not consider the notational scheme applied to ramiform apparatuses to be appropriate for *Panderodus*. Instead, he devised a set of purely descriptive terms (Figs 6.3, 6.4) for all the elements distinguished by earlier workers and also recognized an additional twisted element, which he termed 'tortiform'. Thus, he proposed a quinquemembrate composition for the apparatus. Although these elements may not be comparable with those of ramiform apparatuses, it is conceivable that a set of homologies could be recognized within coniform genera, and, in an attempt to produce a more widely applicable scheme, Armstrong (in press) has applied a modified

| Bergström & Sweet 1966 | Cooper 1975,1976 | Barnes 1977 | Barrick 1977 | Sweet 1979 |
|---|---|---|---|---|
| P. gracilis | costate | narrow | Sa | similiform |
| | | | Sb | asimiliform |
| | | | Sc | arcuatiform |
| P. compressus | simplexiform | wide | M | falciform |
| | | | | tortiform |

| Sweet 1979 | Barnes et al 1979 | Nowlan & Barnes McCracken & Barnes 1981 | Dzik & Drygant 1986 | Armstrong in press |
|---|---|---|---|---|
| similiform | | graciliform | pl | symmetrical p |
| | | | tr | |
| asimiliform | p | | oz | aq |
| | | | sp | |
| arcuatiform | | arcuatiform | ke | r |
| falciform | q | compressiform | ne | sq |
| tortiform | | | hi | tp |

Fig. 6.3—Terminology used by various authors for the elements of the *Panderodus* apparatus.

version of the Type III notation of Barnes *et al.* (1979) to *Panderodus*.

A more simple, trimembrate concept was applied to the type species, *P. unicostatus* Branson and Mehl, by Nowlan and Barnes (1981) and McCracken and Barnes (1981),

who included wide-based (compressiform), costate (graciliform), and asymmetrical costate (arcuatiform) elements. Nowlan and Barnes (1981) identified three distinct apparatus compositions, one typified by the trimembrate *P. unicostatus* and two bimembrate types, and

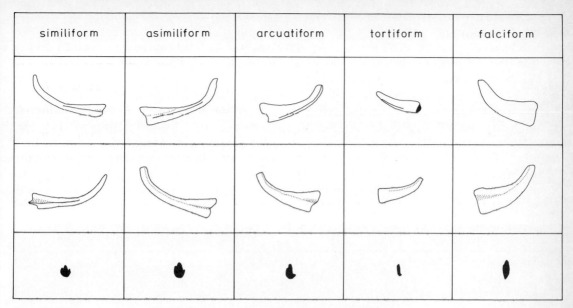

| similiform | asimiliform | arcuatiform | tortiform | falciform |

Panderodus unicostatus ( Branson & Mehl, 1933 )

1 mm

Fig. 6.4—Descriptive terms applied to elements of the *Panderodus* apparatus by Sweet (1979) and followed herein. Sketches show inner lateral face (upper row), outer lateral face (middle row) and transverse section (lower row) of each element type.

suggested that these may provide the basis for a subdivision into three genera. Other authors have suggested that rather more element types may be present in each species. Jeppsson (1983), for example, identified eight to ten element types in each apparatus, although not all could be homologized between species. He included an unpaired, completely symmetrical 'tr' element, although it had a frequency of only 1% in some species. In order to have this element present in all individuals, rather than interpreting it as a juvenile, extreme, or aberrant morphotype, each individual would have to have borne up to one hundred elements. The near-complete fused cluster of *P. unicostatus* from Podolia suggested that this was not the case, and in their description of the specimen Dzik and Drygant (1986) recognized seven element types. They employed the terminology first used by Sweet (1979), but also suggested homologies with the elements of

ramiform apparatuses, which they illustrated by application of the notational scheme devised by Jeppsson (1971). They considered the tr element to be paired and to be subsymmetrical in most individuals. From their clusters, Dzik and Drygant (1986) were able to distinguish two distinct element types within each of Sweet's similiform and asimiliform categories. The similiform types were termed 'pl' and 'tr', with the pl having the more proclined tip. A similar characteristic differentiated the 'sp' and 'oz' elements of asimiliform morphology, with the oz element being proclined to straight and the sp element proclined to erect. Size differences were also recognized between the elements.

We have found the terms proposed by Sweet (1979) to be the most applicable to collections of isolated elements, and employ them here in our discussion of the apparatus structure of *Panderodus*.

## 6.4 ISOLATED CONODONT ELEMENTS FROM THE BRANDON BRIDGE

A 3.1 kg sample from the Brandon Bridge at the Waukesha locality has been processed by Dr R. D. Norby. The conodont collection recovered is moderately diverse and includes the following species: *Dapsilodus obliquicostatus* (Branson and Mehl), *Dentacodina?* sp., *Oulodus? fluegeli* (Walliser), *Oulodus petilus* (Nicoll and Rexroad), *Ozarkodina hadra* (Nicoll and Rexroad), *Ozarkodina polinclinata*

(Nicoll and Rexroad), *Panderodus recurvatus* (Rhodes), *Panderodus unicostatus*, and *Walliserodus* sp. Some of the *Panderodus* elements are preserved as fused clusters.

The 171 specimens of *Panderodus* constitute 72% of the isolated fauna. Comparison of these with topotype material from the Bainbridge Formation of Missouri (Branson and Mehl 1933) shows the dominant species to be *P. unicostatus*, with *P. recurvatus* represented by only 11 specimens. The element types we recognize in *P. unicostatus* conform well with the

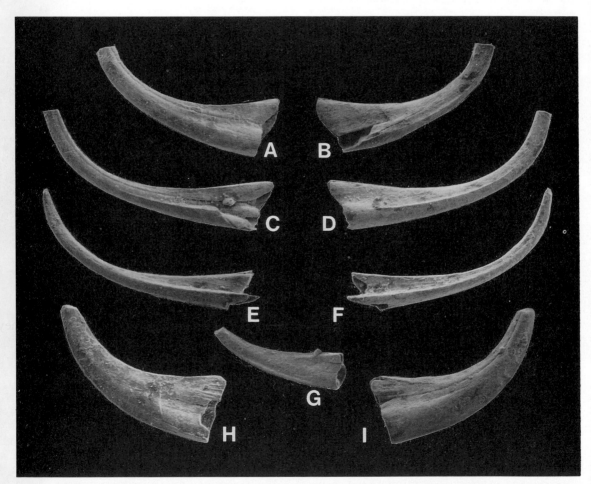

Fig. 6.5—Isolated elements of *Panderodus unicostatus* from the Brandon Bridge of Waukesha, all ×60. A,B, outer and inner lateral views of arcuatiform element, 5797/031. C,D, outer and inner lateral views of asimiliform element, 5797/030. E,F, inner and outer lateral views of similiform element, 5797/029. G, inner lateral view of tortiform element, 5797/032. H, I, inner and outer lateral views of falciform element, 5797/028. All specimens temporarily in the collections of the Micropalaeontology Unit, University of Nottingham, whose numbers are given.

reconstructions of Late Ordovician *Panderodus* species by Sweet (1979), comprising falciform, similiform, asimiliform, arcuatiform, and tortiform elements (Figs 6.3, 6.4, 6.5). These are present in the ratio 2 : 1.4 : 2.2 : 1 : 0.3.

We have been unable to recognize the subtle subdivisions made by Dzik and Drygant (1986) in the isolated elements from Waukesha, as all variation in size and proclination of similiform and asimiliform elements is gradational. It may be that ontogenetic and intraspecific variation in these closely similar element types causes complete overlap in collections of disjunct elements from several individuals, and that the elements could rarely, if ever, be differentiated in such material.

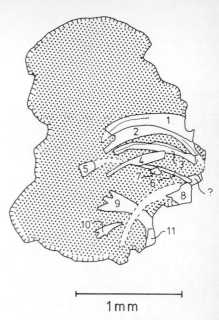

**1 mm**

Fig. 6.6—Conodont assemblage in the *Panderodus unicostatus* animal, UW4001/7a (part); camera-lucida drawing.

## 6.5  THE APPARATUS OF THE WAUKESHA *PANDERODUS* ANIMAL

The elements in the Waukesha animal lie in pairs, with the members opposed across the midline of the apparatus. The long axes of the elements are transverse to the trunk, and the cusps point posteriorly. Identification of the element types is difficult, as several are broken, most are at least partially obscured, and only one lateral face is visible in all cases. For ease of description the elements are numbered sequentially from anterior to posterior (Fig. 6.6).

Only elements 8 and 9 can be identified unequivocally, and these are of falciform morphology (Figs 6.6, 6.7). The upper edges of their bases and their proximal anterior margins are near normal to the midline of the apparatus, and the cusps are proclined to erect. A pair of smaller, fragmentary, and unidentifiable elements (10 and 11) lies to their posterior.

In the central part of the assemblage only small parts of individual elements are preserved or exposed. To the anterior, element 3 exhibits a long base and the proximal part of a proclined cusp. The furrowed, outer, face is uppermost, with no costa visible to the anterior of the furrow. The element is probably asimiliform, as a similiform element would have a costate furrowed face. Elements 1 and 2 are a well-exposed pair and are in contact along most of their length, unlike the falciform pair which are in contact only at the points of maximum recurvature. The furrowed faces of 1 and 2 are uppermost, and no costae are visible. The recurvature is greater than shown by element 3, and the bases are shorter. The distance from the anterior margin to the posterior margin of each element appears to be greater than that in element 3, although the latter is incompletely exposed. The combination of a broad acostate furrowed face and a relatively short base suggests that 1 and 2 are arcuatiform elements.

The elements of the apparatus in the Waukesha animal are similar to the majority of those in the collection of isolated elements from the same locality, indicating that the specimen belongs to *P. unicostatus*.

If we are to use the Waukesha assemblage as a basis for restoring the architecture of the

Fig. 6.7—Conodont assemblage in the *Panderodus unicostatus* animal, UW4001/7b (counterpart): back-scattered scanning electron micrograph of latex cast. Note that all elements display their furrowed outer faces.

## 6.6 APPARATUS ARCHITECTURE OF *PANDERODUS*

The arrangement of elements in the Waukesha assemblage is the result of collapse and flattening of the original three-dimensional apparatus onto a bedding plane. In Chapter 4 of this volume, Aldridge, *et al*. (1987) used a variety of bedding plane assemblages of Carboniferous age to reconstruct the architecture of polygnathacean apparatuses. The Waukesha specimen provides the only comparable evidence for coniform euconodont apparatuses. Nonetheless, the arrangement of elements imposes several constraints on any model.

It is evident from the opposition of the members of element-pairs across the midline of the assemblage that the plane of bilateral symmetry is normal to the bedding. This plane coincides with the midline of the soft parts, and it is apparent that the animal has been dorso-ventrally flattened. Flattening of a number of different architectures in this orientation would result in opposition of the elements. The arrangement could have been radial (resembling the symmetry of the oral disc of lampreys, for example) or in opposing linear arrays, either in a plane or arched (Figs 6.8, 6.9, 6.10).

In an apparatus with radial architecture, the elements could be arranged with their long axes either parallel or transverse to the trunk axis (Fig. 6.8). Flattening of the former, in any orientation, would not produce a linear series of element pairs opposed across the midline. A transverse radial architecture flattened normal to the bedding plane would produce an assemblage in which the elements were superimposed. If the head were tilted, some elements could come to rest in opposition, but others would align with the trunk axis. Hence, normal processes of collapse and flattening of a radial architecture at a high angle to bedding could not produce the Waukesha assemblage. Indeed, the preserved configuration of the elements, spaced along the axis of the specimen in two linear series, indicates that the apparatus, whatever the architecture, is

*Panderodus* apparatus, there are several features of the inter-element relationships that must be taken into account:

(a) The elements occur in pairs arranged in linear sequence from anterior to posterior with cusps pointing adaxially and posteriorly.

(b) The falciform elements lie towards the posterior end of the assemblage.

(c) All elements are similarly oriented (all elements on the part have the outer, furrowed, faces uppermost).

(d) A line drawn to connect the proximal ends of the elements in each half-apparatus runs roughly parallel to the midline. Elements 5 and 10, however, are somewhat offset.

(e) The cusp tips of the elements in each half-apparatus extend approximately to the basal margins of their counterparts.

(f) Although there is variation in element size, it is not gradational from anterior to posterior (*contra* Dzik and Drygant 1986).

(g) Elements at the anterior of the assemblage are more closely spaced than those at the posterior.

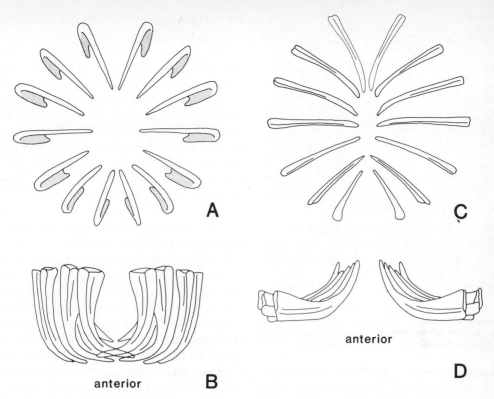

Fig. 6.8—Hypothetical radial architecture for the *Panderodus* apparatus. A, with elements arranged with their long axes parallel to the trunk axis, anterior view. B, the assemblage that would be produced by direct dorso-ventral flattening of A. C, with elements arranged with their long axes transverse to the trunk axis, anterior view. D, the assemblage that would be produced by direct dorso-ventral flattening of C.

unlikely to have been oriented at a high angle to bedding as this would have resulted in stacking or close overlap of the elements. If the head and trunk are flattened in the same orientation to bedding (i.e. dorso-ventrally), then the apparatus in life cannot have been vertical with respect to the trunk.

The alignment of the elements (constraints d and e above) suggests that they were all situated more or less equidistant from the midline. There is no evidence of a horizontal radial arrangement. The position of the elements indicates that they were set in rows parallel to the axis of the trunk (Figs 6.9, 6.10), although

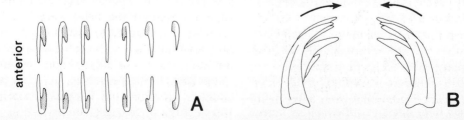

Fig. 6.9—Hypothetical linear architecture for the *Panderodus* apparatus, with the bases of the elements in a plane. A, ventral or dorsal view. B, anterior view; elements illustrated in withdrawn position, to occlude they would be rotated together in the direction shown by the arrows. The Waukesha assemblage is preserved in the occluded position.

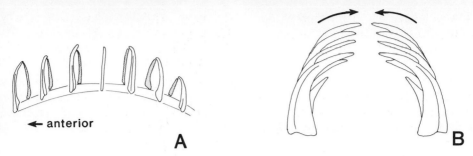

Fig. 6.10—Hypothetical linear architecture for the *Panderodus* apparatus, with the bases of the elements on an arch. A, lateral view; element tips directed inwards and either dorsally or ventrally. B, anterior view, arrows show direction of occlusion.

the plane of the apparatus may not have been horizontal, but could have been tilted. The opposing element rows may have been flat or arched.

Further constraints are introduced by the spacing of the elements in the assemblage, since those in the anterior half are more crowded than those in the posterior. However, this feature cannot be used with confidence on the basis of a single specimen. It may indicate that the spacing was not constant in life. If, on the other hand, the spacing were constant, then the configuration in the Waukesha specimen must reflect the attitude of the head during flattening. An evenly-spaced, planar, linear array would not produce this pattern, no matter what the orientation of the head of the animal relative to the bedding plane. If an arched array, however, were flattened with the head tilted anteriorly, the elements on one limb of the arch would fall more closely together than those on the other. A similar effect would be produced by dorso-ventral flexure of a planar apparatus. (It is pertinent to note that a degree of tilting may also be required to bring the elements of an arched array into a consistent facing direction. Direct dorso-ventral flattening of an arch could result in the preservation of the elements of one limb with their inner faces uppermost, while those of the other limb displayed their outer faces.)

In summary, the assemblage in the Waukesha animal suggests that the apparatus of *P. unicostatus* comprised two opposed halves, operating bilaterally, and arranged as linear arrays either:

(a)  with the anterior elements more closely spaced than the posterior, or

(b)  with the elements evenly spaced and the arrays arched or flexed upwards or downwards in lateral view. An arched arrangement is essentially that deduced by Dzik and Drygant (1986) from the study of fused clusters.

## 6.7   EVIDENCE FROM FUSED CLUSTERS

Several clusters of coniform euconodont elements have been described in which the elements of a half-apparatus are fused laterally. The most complete example is of *Besselodus arcticus*, which comprises six laterally costate (distacodontiform) elements in series, with a geniculate (oistodontiform) element at one end (Aldridge 1982). The close alignment favours a linear architecture, with the geniculate element, by analogy with the Waukesha *Panderodus*, at either the anterior or posterior. A similarly arranged, though more poorly preserved, cluster of *Cordylodus* described by Repetski (1980) possesses three laterally juxtaposed elements together with the cusp tips of up to three opposing elements.

Barnes (1967) and Nowlan (1979) described clusters of laterally fused elements of *Belodina*, a panderodontid, superficially similar to those above. However, each of their clusters shows a different ordering of the elements, and, as concluded by Nowlan (1979), they are probably faecal.

The first clusters of *Panderodus* were recorded by Pollock (1969), but these contain too few elements to permit interpretation. The cluster of *P. unicostatus* described by Dzik and Drygant (1986) is, however, almost complete, and the elements show some of the relationships observed in the Waukesha assemblage. The outer lateral faces of all the elements, for example, face the same way. The Podolian cluster differs significantly, however, in not revealing any direct evidence for two opposed rows of elements, although this is the architecture that Dzik and Drygant (1986) derived. The elements show four principal orientations, with only the falciform ('ne') elements retaining a convincing paired relationship. Furthermore, the sequence of elements on the sinistral side of the assemblage differs from that on the dextral. Lateral flattening of a linear or arched architecture could produce a more irregular assemblage than dorso-ventral collapse, with elements from the two half-apparatuses intermingled, but would be unlikely to significantly alter the sequence of elements on either side. We agree with Dzik and Drygant (1986) that this cluster has been affected by a degree of post-mortem disorganization.

The sequence of elements deduced by Dzik and Drygant (1986) is not in agreement with the data provided by the Waukesha assemblage. They postulated that the size gradient of the elements corresponded to the original order in the apparatus, with the smallest at the posterior and the largest, falciform, elements at the anterior. In the Waukesha specimen large ?arcuatiform elements are at the anterior, but the larger falciform elements (Fig. 6.6, elements 8 and 9) are situated towards the posterior. Although we cannot determine the complete sequence of elements in the assemblage, it does not appear to be one of simple size gradation.

The best evidence for the number of elements in coniform apparatuses is provided by the clusters of *Besselodus* and *Panderodus*. The half-apparatus constituting the *Besselodus* cluster comprises seven elements (Aldridge 1982). In the Podolian *Panderodus*, Dzik and Drygant (1986) recorded six paired elements and a seventh, the tortiform ('hi'), for which the opposing member is missing. It is therefore likely that both these genera possessed apparatuses comprising seven pairs of elements. There is nothing in the Waukesha assemblage to refute this conclusion.

## 6.8  BASAL SUPPORT IN CONIFORM EUCONODONT ELEMENTS

Several authors have suggested that conodont elements were mounted on a supporting structure (e.g. Kirk 1929, Stewart and Sweet 1956, Schwab 1965, Lindström and Ziegler 1971, Jeppsson 1979). A cartilaginous framework was envisaged by Lindström and Ziegler (1971), who suggested that the shrunken appearance of some basal fillings may be due to the post-mortem contraction of a weakly mineralized cartilage-like tissue. Jeppsson (1979) compared the gross morphology of the crown of several conodont elements with the shapes of fish teeth, and pointed out that pike teeth attached to a basal support possess flared bases, similar to those displayed by many coniform conodont elements, including those of *Panderodus*. Unsupported teeth, in contrast, have parallel-sided bases. If conodont elements were supported, the contact between mineralized basal filling and unmineralized support may have been transitional, and the degree of mineralization may have varied between taxa. Jeppsson (1979) extended this model to ramiform and pectiniform elements, concluding that their morphological styles are also in better agreement with attachment to a skeletal framework rather than directly to muscles. The apparatus of *Icriodus*, which contains numerous small coniform elements (Nicoll 1982), lends some support to this contention. Nicoll

(1982) suggested that each of these conical elements might be homologous with a single denticle on the process of a ramiform element in other apparatuses. If this is the case, it would be difficult to envisage how they could function without some form of contiguous basal support.

For coniform elements such as those of *Panderodus*, basal support would provide a more efficient mechanism for bringing the two half-apparatuses into occlusion than would the operation of fourteen individual muscularized elements. Some histological attributes of early euconodonts may provide evidence of the existence of a basal support, although much further work is required. Spherulitic structures have been recorded in the basal fillings of several Ordovician taxa, including *Panderodus* (Barnes *et al*. 1973) and *Coleodus* (Barskov *et al*. 1982). Similar structures are illustrated in *Cordylodus* by Szaniawski (1987, pl. 2.3) in Chapter 2 of this volume. A tissue of similar appearance, globular calcified cartilage, is found in the earliest heterostracans (Denison 1967, Halstead 1969) and in placoderms and elasmobranchs (Ørvig 1951).

Another feature worthy of further attention is the extensive development of the basal body in representatives of the Coleodontidae. Moskalenko (1976) and Barskov *et al*. (1982) figured specimens of *Coleodus* in which pectiniform crowns are mounted on large V-shaped basal bodies composed of a laminated basal cone and a spherulitic basal filling. In *Archeognathus* the crown is developed as a series of bowed coniform denticles connected by strips of crown tissue and a substantial basal body (Klapper and Bergström 1984). This basal body is about 5 mm long and of about the same height as the denticles. It is in the form of a bowed bar with a lateral wing-like process at the anterior. The microstructure is similar to that of typical basal cone tissue, but is more porous (Klapper and Bergström 1984, fig. 10). Each *Archeognathus* unit has generally been considered to be an individual element, but the size is much greater than is normal in Ordovi-

cian conodonts. It is possible that *Archeognathus* represents a rare example in which complete mineralization of the supporting structure of a half-apparatus occurred, and that each 'denticle' is equivalent to an individual element. Klapper and Bergström (1984, p. 972) considered this possibility, but felt it unlikely that clusters of the type shown by *Besselodus* and *Panderodus* could be produced from discrete coniform units arranged in a row as in *Archeognathus*.

## 6.9   IMPLICATIONS FOR AFFINITY

On the basis of the soft-tissue and apparatus structures preserved in the Carboniferous conodont animals from Granton, Aldridge *et al*. (1986) regarded the Conodonta as a group of jawless craniates. The poor preservation of the soft parts in the Waukesha specimen provides little additional evidence of detailed soft-tissue morphology, but the dimensions of the body suggest that *Panderodus* was not as elongate as the Granton polygnathaceans. The preserved symmetry of the Waukesha assemblage indicates that the specimen is flattened in dorso–ventral aspect. If this orientation to bedding reflects a plane of flattening in life it indicates that *Panderodus* may have differed in cross-section from the polygnathaceans which were laterally flattened in life.

The architecture we have reconstructed for the *Panderodus* apparatus is consistent with the postulated relationship between conodonts and pre-gnathostome craniates (Aldridge *et al*. 1986). Similar bilaterally operative, but unmineralized, jaw structures are present in the myxinoids (Dawson 1963, Janvier 1981) and petromyzontids (Janvier 1981, Yalden 1985). The oral plates of heterostracans may also have been operated by comparable musculature (Janvier 1981). Coniform euconodont assemblages provide no evidence of structures like the oral disc of petromyzontids or the oral plates of heterostracans. The opposing half-apparatuses recognised in *Pan-*

*derodus* and *Besselodus* bear most resemblance to the lingual apparatus of myxinoids.

## ACKNOWLEDGEMENTS

Our work on conodont palaeobiology is financed by N.E.R.C. Research Grant GR3/5105. This chapter is related to a project on the Waukesha biota coordinated by Dr Don Mikulic, Illinois Geological Survey, to whom we are grateful for advice and encouragement. The Waukesha specimen was loaned by Dr K. Westphal, Geology Museum, University of Wisconsin—Madison. Dr Rod Norby, Illinois Geological Survey, supplied the collection of isolated conodont elements from Waukesha, and Dr D. J. Kennedy, Brock University, St Catharines, Ontario, provided topotype material from the Bainbridge of Missouri. We also thank Dr Jerzy Dzik, Academy of Sciences, Warsaw, for providing a prepublication copy of the paper by himself and Dr D. M. Drygant on *Panderodus* clusters from Podolia. The line drawings were drafted by Mrs J. Wilkinson, and photographic assistance was provided by Mr A. Swift.

## REFERENCES

Aldridge, R. J. 1982. A fused cluster of coniform conodont elements from the late Ordovician of Washington Land, western North Greenland. *Palaeontology* **25** 425–430.

Aldridge, R. J., Briggs, D. E. G., Clarkson, E. N. K., and Smith, M. P. 1986. The affinities of conodonts—new evidence from the Carboniferous of Edinburgh, Scotland. *Lethaia* **19** 279–291.

Aldridge, R. J., Smith, M. P., Briggs, D. E. G., and Norby, R. D. 1987. The architecture and function of Carboniferous polygnathacean conodont apparatuses. In: R. J. Aldridge, (ed.): *Palaeobiology of Conodonts*. Ellis Horwood, Chichester, Sussex, 63–75.

An, Taixiang, *et al.* 1983. *The conodonts of north China and the adjacent regions* (in Chinese, with English abstract). Science Press of China, 223 pp., 33 pls.

Armstrong, H. A. (in press). Conodonts from the early Silurian carbonate platform of North Greenland. *Grønlands Geologiske Undersøgelse Bulletin.*

Barnes, C. R. 1967. A questionable natural conodont assemblage from Middle Ordovician limestone, Ottawa, Canada. *Journal of Paleontology* **41** 1557–1560.

Barnes, C. R. 1977. Ordovician conodonts from the Ship Point and Bad Cache Rapids Formations, Melville Peninsula, southeastern District of Franklin. *Geological Survey of Canada Bulletin* **269** 99–119.

Barnes, C. R., Kennedy, D. J., McCracken, A. D., Nowlan, G. S. and Tarrant, G. A. 1979. The structure and evolution of Ordovician conodont apparatuses. *Lethaia* **12** 125–151.

Barnes, C. R., Sass, D. B., and Poplawski, M. L. S. 1973. Conodont ultrastructure: The family Panderodontidae. *Life Sciences Contributions Royal Ontario Museum* **90** 36 pp.

Barrick, J. E. 1977. Multielement simple-cone conodonts from the Clarita Formation (Silurian), Arbuckle Mountains, Oklahoma. *Geologica et Palaeontologica* **11** 47–68.

Barskov, I. S., Moskalenko, T. A., and Starostina, L. P. 1982. New evidence for the vertebrate nature of the conodontophorids. *Paleontologicheskii Zhurnal*, 1982(1), 82–90.

Bergström, S. M. and Sweet, W. C. 1966. Conodonts from the Lexington Limestone (Middle Ordovician) of Kentucky and its Lateral equivalents in Ohio and Indiana. *Bulletin of American Paleontology* **50** 271–441.

Branson, E. B. and Mehl, M. G. 1933. Conodonts from the Bainbridge (Silurian) of Missouri. *University of Missouri Studies* **8** 39–52.

Briggs, D. E. G., Clarkson, E. N. K., and Aldridge, R. J. 1983. The conodont animal. *Lethaia* **16** 1–14.

Cooper, B. J. 1975. Multielement conodonts from the Brassfield Limestone (Silurian) of southern Ohio. *Journal of Paleontology* **49** 984–1008.

Cooper, B. J. 1976. Multielement conodonts from the St Clair Limestone (Silurian) of southern Illinois. *Journal of Paleontology* **50** 205–217.

Cooper, B. J. 1981. Early Ordovician conodonts from the Horn Valley Siltstone, Central Australia. *Palaeontology* **24** 147–183.

Dawson, J. A. 1963. The oral cavity, the 'jaws' and the horny teeth of *Myxine glutinosa*. In: A. Brodal and R. Fange (eds): *The biology of* Myxine Universitetsforlaget. Oslo, 231–255.

Denison, R. H. 1967. Ordovician vertebrates from western United States. *Fieldiana, Geology* **16** 131–192.

Dzik, J. and Drygant, D. 1986. The apparatus of

panderodontid conodonts. *Lethaia* **19** 133–141.

Halstead, L. B. 1969. Calcified tissue in the earliest vertebrates. *Calcified Tissue Research* **3** 107–124.

Janvier, P. 1981. The phylogeny of the Craniata, with special reference to the significance of fossil 'agnathans'. *Journal of Vertebrate Paleontology* **1** 121–159.

Jeppsson, L. 1971. Element arrangement in conodont apparatuses of *Hindeodella* type and in similar forms. *Lethaia* **4** 101–123.

Jeppsson, L. 1979. Conodont element function. *Lethaia* **12** 153–171.

Jeppsson, L. 1983. Simple cone studies: some provocative thoughts. *Fossils and Strata* **15** 86.

Kirk, S. R. 1929. Conodonts associated with Ordovician fish fauna of Colorado—a preliminary note. *American Journal of Science, series* 5 **18** 493–496.

Klapper, G. and Bergström, S. M. 1984. The enigmatic Middle Ordovician fossil *Archeognathus* and its relations to conodonts and vertebrates. *Journal of Paleontology* **58** 949–976.

Lindström, M. and Ziegler, W. 1971. Feinstrukturelle Untersuchungen an Conodonten.I.Die Überfamilie Panderodontacea. *Geologica et Palaeontologica* **5** 9–33.

McCracken, A. D. and Barnes, C. R. 1981. Conodont biostratigraphy and paleoecology of the Ellis Bay Formation, Anticosti Island, Quebec, with special reference to late Ordovician–early Silurian chronostratigraphy and the systemic boundary. *Geological Survey of Canada Bulletin* **329** 51–134.

Mikulic, D. G., Briggs, D. E. G., and Kluessendorf, J. 1985a. A Silurian soft-bodied biota. *Science* **228** 715–717.

Mikulic, D. G., Briggs, D. E. G., and Kluessendorf, J. 1985b. A new exceptionally preserved biota from the Lower Silurian of Wisconsin, USA. *Philosophical Transactions of the Royal Society of London Series B* **311** 78–85.

Moskalenko, T. A. 1976. Unique conodontophorid finds in the Ordovician deposits of the Irkutsk Amphitheater. *Doklady Akademii Nauk SSSR* **229** 232–234.

Nicoll, R. S. 1982. Multielement composition of the conodont *Icriodus expansus* Branson and Mehl from the Upper Devonian of the Canning Basin, Western Australia. *BMR Journal of Australian Geology and Geophysics* **7** 197–213.

Nowlan, G. S. 1979. Fused clusters of the conodont genus *Belodina* Ethington from the Thumb Mountain Formation, Ellesmere Island, District of Franklin. *Geological Survey of Canada Paper* **79–1A** 213–218.

Nowlan, G. S. and Barnes, C. R. 1981. Late Ordovician conodonts from the Vaureal Formation, Anticosti Island, Quebec. *Geological Survey of Canada Bulletin* **329** 1–49.

Ørvig, T. 1951. Histologic studies of placoderms and fossil elasmobranchs. I: The endoskeleton, with remarks on the hard tissues of lower vertebrates in general. *Arkiv för Zoologi* **2** 321–354.

Pollock, C. A. 1969. Fused Silurian conodont clusters from Indiana. *Journal of Paleontology* **43** 929–935.

Repetski, J. E. 1980. Early Ordovician fused conodont clusters from the western United States. *Abhandlungen der Geologischen Bundesanstalt* **35** 207–209.

Schwab, K. W. 1965. Microstructure of some Middle Ordovician conodonts. *Journal of Paleontology* **39** 590–593.

Smith, M. P. 1985. Ibexian–Whiterockian (Ordovician) conodont palaeontology of East and eastern North Greenland. *Unpublished PhD thesis, University of Nottingham*, 364 pp., 22 pls.

Stewart, G. A., and Sweet, W. C. 1956. Conodonts from the Middle Devonian bone beds of central and west-central Ohio. *Journal of Paleontology* **30** 261–273.

Sweet, W. C. 1979. Late Ordovician conodonts and biostratigraphy of the western Midcontinent Province. *Brigham Young University Geology Studies* **26** part 3, 45–85.

Sweet, W. C. and Schönlaub, H. P. 1975. Conodonts of the genus *Oulodus* Branson and Mehl, 1933. *Geologica et Palaeontologica* **9** 41–59.

Szaniawski, H. 1987. Preliminary structural comparisons of protoconodont, paraconodont and euconodont elements. In: R. J. Aldridge (ed.): *Palaeobiology of Conodonts*. Ellis Horwood, Chichester, Sussex, 35–47.

Yalden, D. W. 1985. Feeding mechanisms as evidence for cyclostome monophyly. *Zoological Journal of the Linnean Society* **84** 291–300.

# 7

# Soft tissue matrix of decalcified pectiniform elements of *Hindeodella confluens* (Conodonta, Silurian)

L. E. Fåhræus and G. E. Fåhræus-van Ree

## ABSTRACT

Pectiniform conodont elements of the Silurian species *Hindeodella confluens* Branson and Mehl from Gogs, Gotland, Sweden, were demineralized, fixed, dehydrated, embedded in paraffin, and sectioned at 6 µm. The sections were stained with haemalum and eosin. With regard to the intensity and localization of the stains the results are indistinguishable from sections obtained with tissue from freshly killed organisms. Examination of the sections reveals variously shaped cells, bundles of fibrous material (possible collagen), collagen, and lumina. Bundles of fibrous material and collagen occur throughout the sections, but constitute less than 25%. This is considerably different from known biomineralized tissue which is totally dominated by collagen.

## 7.1 INTRODUCTION

Any kind of hard tissue incorporating minerals that is found among animals, plants, or protoctistans, is the product of biomineralization. Examples include bone in vertebrates, external shells in molluscs, coccoliths in coccolithophorids, and calcium oxalate deposits in plants. All kinds of biomineralization appear to occur on an organic matrix that is an extra-cellular component of the tissue; that is, it is produced and extruded by the cells (for examples, see Nancollas 1982). Although there is controversy about the actual initiation and process of mineralization, it appears to be generally agreed that cells might induce the mineralization process but are not themselves calcified (Nancollas 1982, Westbroek and De Jong 1983). However, in this contribution we preliminarily report on the first experiment involving decalcification of conodont elements and subsequent histological staining of the freed soft tissue. This experiment has shown that the mineralized tissue of conodonts unequivocally includes cells.

## 7.2 MATERIAL AND METHODS

Pectiniform Pa elements of *Hindeodella confluens* (referred by many authors to *Ozarkodina*, see Fåhræus 1969, pl. 1, fig. 1 for illustration of a typical specimen) collected at Gogs, Gotland, Sweden, were decalcified individually in a commercial solution 'Cal-Ex II' (Fisher Scientific Company, Fair Lawn, New Jersey), which includes a decalcifer (trichloroacetic acid), a fixing agent (formaldehyde), a mordant, and a surfactant. Initial attempts at using decalcifier and fixing agent separately were unsuccessful, as the freed tissue disintegrated almost immediately upon exposure. Following dissolution and fixation, which took approximately 15–20 minutes depending on the size of the element, the conodont tissue was dehydrated and embedded in paraffin. The embedded tissue was sectioned (6 $\mu$m) and the sections were stained with haemalum and eosin.

## 7.3  RESULTS

The following descriptions are based on observations obtained from the soft tissue residue secured from one conodont element. With regard to the intensity and localization of the stains, the results of the staining of the sections are indistinguishable from sections obtained with tissue from freshly killed organisms.

Under the light microscope the sections consist of cells, fibrous material, and lumina. The lumina (varying in estimated cross-section from 5–40 $\mu$m) are either round, irregularly shaped, or represented as long canals. The round lumina are smoothly lined by cells with elongated nuclei and weakly eosinophilic cytoplasm. Some of these lumina are outlined by dark pigment. The irregularly shaped lumina are sometimes partially outlined by what appears to be weakly-stained to practically unstained cytoplasm with occasional round nuclei which are surrounded by flat to round nucleoli. The long canals, which sometimes appear to be branched, are frequently lined by elongated cells with very small nuclei. The fibrous material, which is rather sparsely dispersed among the cells, is non-anastomosing, occurs as bundles, and, in most cases, appears to lack nuclei. Some of it is birefringent and pale rose in colour. Other fibrous material is strongly basophilic and densely packed, and appears to be associated with elongated, generally sinuous cells.

Other cells vary in shape and size: there are relatively large cells with nuclei of 3–5 $\mu$m and of longest diameter 15–20 $\mu$m, and smaller cells with many small nucleoli and strongly eosinophilic cytoplasm. The shape of these cells is very variable: angular, round, or elongated. In close proximity to these large cells there are cells with round to elongated nuclei which are distinctly smaller than the nuclei of the larger cells. These cells also have many nucleoli and less or no eosinophilic cytoplasm (Pl. 7.1).

## 7.4  INTERPRETATION AND DISCUSSION

The most remarkable result of this study is the fact that tissue 400 Ma old can remain histochemically intact with regard to its capability to react with biological stains. It has been generally promulgated that cell structures and collagen fibrils, forming the matrix in bone and other products of biomineralization, disintegrate within a short time upon death, and only their degradation products in the form of amino acids remain. Earlier studies have shown that conodonts as old as 470 Ma retain detectable amino acids (Pietzner *et al*. 1968).

The obvious question in a study like this is: are we dealing with contaminants? There

Plate 7.1
Fig. 1—Unoriented section of pectiniform element soft-tissue matrix showing different kinds of cells, extra-cellular substance, and lumina. Haemalum and eosin staining. Magnification ×130.
Fig. 2—Detail of pectiniform element soft-tissue matrix showing a canal with flat cell lining (small arrow), branched canal (large arrow), large cell (open arrow), and fibrous material (long arrow). Haemalum and eosin staining. Magnification ×530.
Fig. 3—Detail of pectiniform element soft-tissue matrix showing a canal outlined by pigment (small arrow) and large cells. Haemalum and eosin staining. Magnification ×530.
Fig. 4—Detail of pectiniform element soft-tissue matrix showing fibrous (? collagen) material (small arrow) and collagen (large arrow). Haemalum and eosin staining. Magnification ×530.

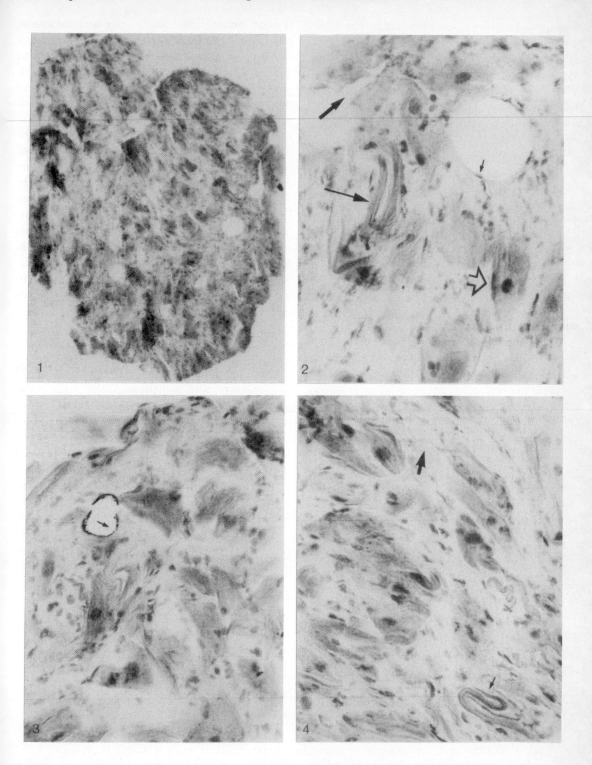

appear to be two possible sources for contaminants: (1) Recent boring or other organisms occupying existing voids, and (2) contaminants introduced during processing.

The first possibility can be ruled out simply because of scale. The piece of tissue sectioned for this study would have occupied approximately two-thirds of the length of the decalcified element, and, furthermore, no indications of boring or other organisms were observed on the untreated specimen. With regard to the second possibility our primary reasons for deciding that we are not dealing with contaminants (and we certainly worried about the possibility) are remnants of undissolved mineralization in the tissue, detected through typical calcite birefringence under crossed nicols in several of our sections. All preparations have been made with material and equipment never previously used for histological preparations (or any other purpose), in a laboratory never before used for any kind of histological preparation. In addition, we have shown our sections to numerous histologists, none of whom has been able to name a modern type of tissue with similar histomorphologies. We therefore feel reasonably safe in saying that our material actually represents the decalcified conodont element matrix.

We believe that the preservation of soft tissue and the at least partially intact histochemistry of our 400 Ma old conodont elements may be explained by three factors. Firstly, conodont elements have considerably less porosity than bone, a fact which should have resulted in less external influence on the internal structures once the life-supporting supplies were cut off. Secondly, the colour of the conodont elements used in this study shows that they were never exposed to temperatures higher than 50–80°C (based on the work of Epstein *et al.* 1977); thus no thermal degradation of organic material will have occurred. Thirdly, the conodont elements used in this study were probably never sub-aerially exposed and thus never subject to weathering.

The lumina represent the only features that can be directly identified with known structures in the intact conodont element. Unaltered elements are usually light amber in colour. However, in certain areas of the element, particularly in the cusp and denticles and along their bases in compound forms, tiny vesicles or cancellated features can be observed under the light microscope as white opaque structures. These structures have been referred to as 'white matter' (see Lindström and Ziegler 1981, p. W48). Under the scanning electron microscope (SEM) these structures show up as regular to irregular voids, varying in size from $0.1–0.5$ $\mu$m, with a stated maximum diameter of 1 $\mu$m (Lindström and Ziegler 1981). These voids most probably represent transverse sections of vessels, which in some cases appear to be branched; the voids in the particular conodont elements used for our study may have a diameter of 30 $\mu$m or more. Such voids are very common in the untreated specimens, and, in our decalcified sections, transverse and more or less oblique sections of lumina are very common. It is therefore concluded that the conodont tissue was highly vascularized. The discrepancy between the size of the lumina in our specimens and that reported for voids by Lindström and Ziegler (1981) may be due to whereabouts on the element the voids (lumina) were observed; i.e. under the SEM such voids have been observed close to the tip of the cusp, whereas in our study more basally located voids were observed. It is also quite likely that the size of voids varies with the type and size of elements; the Pa elements used in this study are quite large, and SEM examination of voids from this type of element has not been reported. The elongated cells lining some of the lumina probably represent a type of endothelium. Our study indicates that 'white matter' is the optical representation of the vascularized tissue of the conodont element. The extensive vascularization implies a very active tissue demanding lots of nutrients.

At least some of the fibrous tissue, because of its structure, birefringence, and lack of nuclei, is considered to be collagen. There are bundles of collagen with small fibroblasts (with large nuclei) interspersed between the larger cells.

We are uncertain about the nature of the distinctly larger cells, and refer to them simply as 'mesenchyma' cells. There are also other cell morphologies, clearly identifiable in Plate 7.1, that we prefer at present not to describe in terms implying an identity with cell types recognized in modern tissue.

We wish to stress that the major difference between the decalcified matrix of the conodont elements and any known type of biomineralization is that the conodont matrix includes cells. Proportionally only about 20–25% of our sections is represented by collagen. In other types of biomineralized tissue collagen and some other extra-cellular products (macromolecules) account for the entire matrix.

Conodont elements compositionally consist of carbonate-apatite, which sometimes incorporates minor amounts of fluorine (Lindström and Ziegler 1981). Mineralogically they are thus closest to francolite, a product of biomineralization that is known to occur among inarticulate brachiopods, certain molluscs, and chordates. However, the decalcified conodont tissue is not bone or dental tissue as we presently know it. Bone is either totally non-cellular with an extra-cellular matrix largely consisting of collagen fibrils or, in cellular bone, it is dominated by bundles of collagen with the occasional 'trapped' osteocyte. Dentine and enamel, in teeth, are totally different in their matrix structure from the decalcified conodont elements. The type of tissue encountered in this study is also in no way similar to that found among extant invertebrates with apatitic biomineralization such as inarticulate brachiopods and molluscs.

On the basis of known types of biomineralization (e.g. Westbroek and De Jong 1983, Watabe and Wilbur 1976) we exclude any direct phylogenetic relationship between conodonts and inarticulate brachiopods or known types of mollusc. With regard to a phylogenetic relationship with the chordates we are more cautious, since studies of cellular mineralized tissue of early chordates have exclusively been based on the morphologies of cavities (presumed cell lacunae) as seen in thin sections of armour, scales, and teeth. Open spaces in those of our sections that are poorly decalcified show that cavities in the undissolved tissue do not necessarily correspond to cell outlines. Until stained tissue sections of early Palaeozoic vertebrate tissue (e.g. ostracoderms) have been produced we do not rule out the possibility that conodont elements may be the result of an early type of vertebrate biomineralization. Neither, of course, do we rule out the possibility that the conodont elements represent an entirely unique approach to biomineralization. We welcome ideas and suggestions from readers of this preliminary report.

## ACKNOWLEDGEMENT

L.E.F. wishes to acknowledge continued financial support from the Natural Sciences and Engineering Research Council of Canada for the study of the Palaeobiology of conodonts.

## REFERENCES

Epstein, A. G., Epstein, J. B., and Harris, L. D., 1977. Conodont color alteration—an index to organic metamorphism. *US Geological Survey Professional Paper* **995** 27 pp.

Fåhræus, L. E., 1969. Conodont zones in the Ludlovian of Gotland and a correlation with Great Britain. *Sveriges Geologiska Undersökning*, ser. C, Nr. **639** 33 pp., 2 pls.

Lindström, M. and Ziegler, W. 1981. Surface micro-ornamentation and observations on internal composition. In: R. A. Robison (ed.): *Treatise on Invertebrate Paleontology, Part W, Supplement 2, Conodonta*, Geological Society of America and University of Kansas Press, Lawrence, Kansas, W41–W52.

Nancollas, G. H. (ed.) 1982. *Biological Mineralization and Demineralization*. Springer-Verlag Berlin, Heidelberg, New York, 415 pp.

Pietzner, H., Vahl, J., Werner, H., and Ziegler, W.

1968. Zur chemischen Zusammensetzung und Mikro-morphologie der Conodonten. *Palaeontographica* **128** 115–152, pls. 18–27.

Watabe, N., and Wilbur, K. M., (eds.) 1976. *The mechanisms of mineralization in the invertebrates and plants*. University of South Carolina Press, Columbia, South Carolina, 461 pp.

Westbroek, P. and De Jong, E. W. (eds.) 1983. *Biomineralization and biological metal accumulation*. D. Reidel Publishing Company, Dordrecht/Holland, Boston/USA, London/England, 532 pp.

# 8

# An abrupt change in conodont faunas in the Lower Ordovician of the Midcontinent Province

R. L. Ethington, K. M. Engel, and K. L. Elliott

## ABSTRACT

Detailed sampling in the Lower Ordovician of Utah and Oklahoma has shown that the conodonts of Fauna C of Ethington and Clark (1971) are replaced by those of Fauna D within an interval of less than a metre. Fauna D is introduced with a few coniform species and diversifies up-section. The faunal replacement occurs within a sequence of strata that gives evidence of progressive shoaling during deposition. The transition from rocks with Fauna C to those with Fauna D is suggestive of the relations at the boundaries of biomeres that have been defined from trilobite occurrences.

## 8.1  INTRODUCTION

In one of the earliest systematic studies of conodonts from the Lower Ordovician of the North American Platform, Furnish (1938) noted conspicuous differences between the faunas of the Oneota Formation and those of the overlying Shakopee Formation in the Upper Mississippi Valley region. He found few species that he reported to persist through this stratigraphical interval, and observed that even these generally show differences as they are developed in the two formations. Subsequently, these two faunal assemblages have been shown to be widely distributed in shallow-/warm-water deposits in North America, and components of them have been recognized from the Siberian Platform as well (Moskalenko 1967, Abaimova 1975). They are not present, however, in the thoroughly studied coeval rocks of deep-/cool-water origin in Baltoscandia (Lindström 1955, Sergeeva 1966, Viira 1974).

By 1970 the geographical and stratigraphical distributions of these conodonts had been established sufficiently to permit Ethington and Clark (1971) to consider them as characteristic of two of the five distinct faunas that they reported to exist in homotaxial sequence in the Lower Orodvician of North America. Faunas C and D, as they designated these two faunal assemblages, have become standard frames of biostratigraphical reference in subsequent discussions of Lower Ordovician rocks in which they occur. Typical species of Faunas C and D are illustrated on Plate 8.1.

The top of the stratigraphical range of Fauna C was reported by Ethington and Clark (1982) to coincide with the top of the House Formation in western Utah. They found the lower 33.5 m of the overlying Fillmore Formation to contain sparse conodonts of low diversity. Typical species that they earlier had considered to be characteristic of Fauna D such as *Glyptoconus quadraplicatus* (Branson and Mehl), *Eucharodus parallelus* (Branson and Mehl), and *Scolopodus rex* Lindström (= *S. cornutiformis* of Ethington and Clark 1971) were not found in this interval, but they become increasingly abundant above it and have long stratigraphical ranges higher in the Fillmore. Ethington and Clark believed that only two species, *E. parallelus* and '*Scolopodus*' *sulcatus* Furnish, persist across the formational and faunal boundaries at the top of the House; they concluded that no close phylogenetic affinity exists between the other conodonts of the

upper part of the House Formation and those in the lower Fillmore. The abrupt replacement of Fauna C by unrelated conodonts was viewed as one of the most profound changes in the early history of conodonts, and one whose stratigraphical expression offers an opportunity for very firm correlations within the Lower Ordovician of the Midcontinent Province.

This generalization was based on only two sections, the one in western Utah that Ethington and Clark had studied in detail, and another in western Texas (Repetski 1982) in which closely spaced sampling had been undertaken. Elsewhere, rocks containing Faunas C and D have been reported at a reconnaissance level only (e.g. Sando 1958), and have been

---

Plate 8.1—Figures 1–5, 11 illustrate typical conodont elements of Fauna D, figures 8–10 represent the low-diversity fauna that is introduced following extinction of Fauna C, and figures 6, 7, 11–21 are of typical conodont elements of Fauna C.

Fig. 1—*Diaphorodus delicatus* (Branson and Mehl), ×100; Highway 77 Section, 27 m above base of West Spring Creek Formation.

Fig. 2—*Ulrichodina wisconsinensis* Furnish, ×70, Highway 77 Section, Cool Creek Formation, 235 m below top.

Fig. 3—*Glyptoconus quadraplicatus* (Branson and Mehl), ×65, Highway 77 Section, Cool Creek Formation, 235 m below top.

Fig. 4—*Eucharodus parallelus* (Branson and Mehl), ×40, Highway 77 Section, Cool Creek Formation, 235 m below top.

Fig. 5—*Scolopodus rex* Lindström, ×60, Highway 77 Section, Kindblade Formation, 103 m above base.

Fig. 6—'*Oistodus*' *triangularis* Furnish, ×100, Highway 77 Section, McKenzie Hill Formation, 9 m below top.

Fig. 7—?*Acanthodus uncinatus* Furnish, ×40, Chandler Creek Section, McKenzie Hill Formation, 0.3 m below top.

Fig. 8—?*Oneotodus* sp., ×180, Highway 77 Section, Cool Creek Formation, 0.9 m above base.

Fig. 9—?*Oneotodus* sp., ×80, Highway 77 Section, Cool Creek Formation, 0.9 m above base.

Fig. 10—?*Oneotodus* sp., ×180, Highway 77 Section, Cool Creek Formation, 0.9 m above base.

Fig. 11—*Macerodus dianae* Fåhræus and Nowlan, ×140, Highway 77 Section, Cool Creek Formation, 235 m below top.

Fig. 12—*Cordylodus angulatus* Pander, ×150, Highway 77 Section, McKenzie Hill Formation, 7.5 m below top.

Fig. 13—*Chosonodina herfurthi* Müller, ×80, Highway 77 Section, McKenzie Hill Formation, 9 m below top.

Fig. 14—*Rossodus manitouensis* Repetski and Ethington, ×100, Chandler Creek Section, McKenzie Hill Formation, 7.5 m below top.

Fig. 15–'*Paltodus*' *bassleri* Furnish, ×125, Manitou Formation, Missouri Gulch Section, Colorado.

Fig. 16—*Acanthodus lineatus* (Furnish), ×100, Chandler Creek Section, Cool Creek Formation, 0.9 m above base.

Fig. 17—*Clavohamulus densus* Furnish, ×190, Highway 77 Section, McKenzie Hill Formation, 3 m below top.

Fig. 18—'*Acontiodus*' *iowensis* Furnish, ×90, Chandler Creek Section, Cool Creek Formation, 1.5 m above base.

Fig. 19—'*Acodus*' *oneotensis* Furnish, ×90, Chandler Creek Section, McKenzie Hill Formation, 1.8 m below top.

Fig. 20—*Loxodus bransoni* Furnish, ×70, Highway 77 Section, McKenzie Hill Formation, 9 m below top.

Fig. 21—*Oneotodus datsonensis* Druce and Jones, ×180, Highway 77 Section, McKenzie Hill Formation, 0.3 m below top.

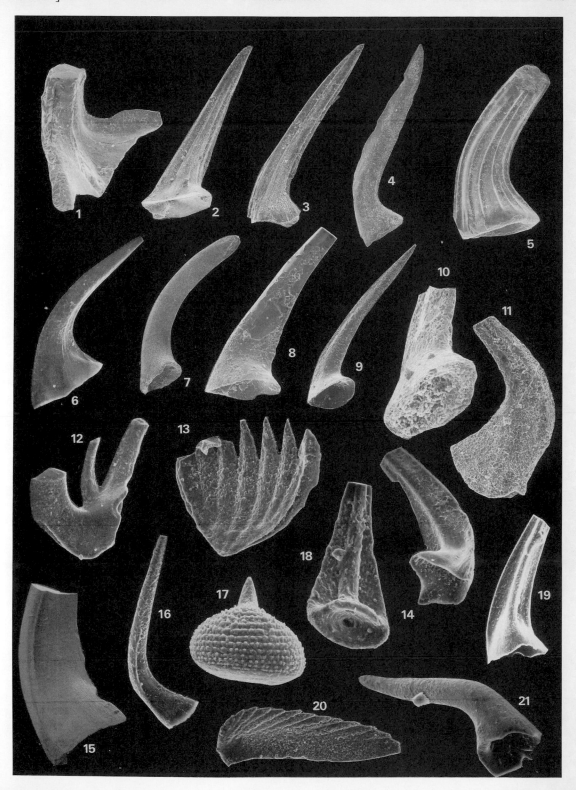

collected at broadly spaced intervals (Nowlan 1985), or involve composite sections (Furnish 1938); all of these reports provide low biostratigraphical resolution that would not detect an abrupt faunal replacement. Ethington and Clark found very low abundances of conodonts in the productive samples from lower Fillmore, so that the sudden introduction of Fauna D following disappearance of Fauna C that they reported could be an artifact of poor recovery in the critical interval in western Utah.

To clarify the biostratigraphical distributions of these conodont faunas, we sampled in detail across the C–D boundary interval in western Utah. Larger and more closely spaced samples than those collected by Ethington and Clark (1982) were assembled with the intention of either refuting the nearly mutually exclusive occurrences that they described or of adding details, particularly for Fauna D, if those occurrences were substantiated. Conodonts characteristic of Faunas C and D also occur in the Arbuckle Group in Southern Oklahoma (Mound 1968). We collected through the boundary interval for these two faunas at two localities there in order to obtain data for comparison with the information from the Utah succession. The conodonts recovered from these collections are reported and evaluated in the discussion that follows.

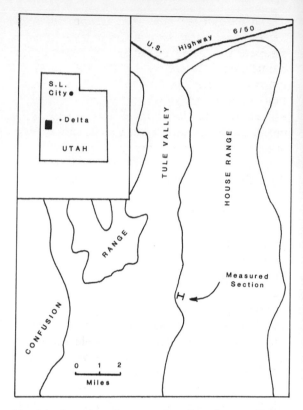

Fig. 8.1—Location of measured section of upper House and lower Fillmore formation, Ibex area, western Utah.

## 8.2   IBEX AREA, UTAH

### (a) Section location

The sections sampled by Ethington and Clark (1982) are in the steep, west-facing slope of the House Range. This north–south trending, block-faulted region exposes the House and Fillmore formations in a shallow syncline whose axis is approximately normal to the length of the range. The contact between the formations is high on the slope to the south, but descends on the southern limb of the syncline and disappears beneath the alluvial floor

of Tule Valley near the synclinal axis, rising again on the northern limb beyond U.S. Highway 6–50 (see Fig. 8.1). The locality selected for study is approximately 19.5 km south of the highway and about 0.8 km east of the unimproved road that extends down the east side of Tule Valley. Here (NW1/4, Sec. 31, T22S, R13W; The Barn, Utah 15 min. Quadrangle) a southwesterly trending canyon dissects the face of the range. Our measured section, which is in the general vicinity of Section B of Hintze (1951) and of Ethington and Clark (1982), is in the northwesterly trending tributary that joins the main valley near its mouth. The steep slope of the canyon floor and the general absence of soil and vegetation in this arid region provide complete exposures of the sequence of beds under consideration.

The base of the measured section, which is located where the floor of the valley begins to steepen, is marked by the letters KME in red paint. Samples were collected at 0.61 m (2 ft) intervals through a 24.4 m (80 ft) succession that includes the top 11.3 m of the House Limestone and the lower 13.1 m of the Fillmore. Sampled horizons are indicated on the outcrop by prominent red numbers.

## (b) Lithostratigraphy

In the sampled interval, the House Limestone consists of thin to thick beds of limestone, principally lime mudstones but with some beds of lime grainstone and of lime wackestone. These ledges alternate with intervals of silty limestone (up to 1.5 m) that weather to form re-entrants in the slope. The limestones typically are medium dark grey, but the silty interbeds display brownish hues. Ripple marks of low amplitude and wavelength as well as cross-laminations are sporadically distributed through this uppermost part of the formation. Black chert in lenses of the order of 5 cm thick and up to a metre in length occur in 5–10% of the beds in the sequence. Flat-pebble conglomerates with thin, discoidal intraclasts occur at several places in the section but are not as common as in the overlying Fillmore Formation. A generalized columnar section through the sampled interval is shown in Fig. 8.2.

The lithology of the sampled part of the Fillmore Formation does not differ significantly from that of the underlying House Limestone. Those beds that have a relatively low content of detrital material protrude as ledges whereas silty strata form re-entrants and slopes. The frequency of occurrence of flat-pebble conglomerates increases up-section as does the thickness of individual units of this type. Grainstones in which the grains consist of worn fragments of trilobites and of brachiopods are more common in this sequence than in the House.

We have not observed an abrupt change in lithofacies at any level in the sampled interval.

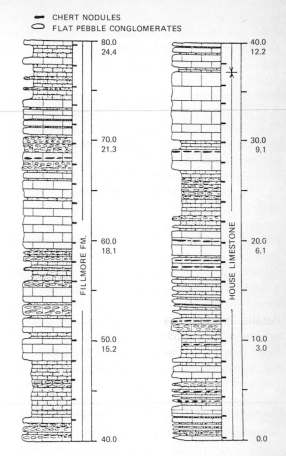

Fig. 8.2—Columnar section at Ibex; sampled horizons indicated by tick-marks to right of section; section calibrated in feet (above) and metres (below).

The general increase in frequency of occurrence of flat-pebble conglomerates accompanied by an increase in abundance of grainstones at the expense of mudstones suggests a progressively shallowing depositional regime. The section was examined carefully for the presence of stratigraphical discontinuities which might account for the sudden termination of the range of the older fauna. No obvious discordance was found, and corrosion zones or hard grounds were not detected in petrographic examination of hand specimens. A discontinuity may be present in one of the re-entrants in the lower Fillmore, but if so it is a feature of low relief. Lehi Hintze, who has done extensive regional studies in western Utah (1951, 1973), has not reported physical evidence for unconformity in the upper House-lower Fillmore part of the section.

### (c) Conodont biostratigraphy

A kilogram of rock from each of the samples was processed; only two of the samples (those at 0.61 m and at 17.1 m) were barren. The 39 productive samples produced 3964 elements, but the abundance varied markedly. Samples collected through the top of the House Limestone yielded an average of 154 elements per sample; maximum yield (sample 10) was 573 elements, and only seven samples (1, 3, 5, 8, 13, 14, 16) produced fewer than 50 elements. In contrast, the average yield in samples from lower Fillmore was 57 elements with a maximum of 206 (sample 25) and with 11 samples having fewer than 15 elements. the most abundant samples in both intervals were collected from ledges of lime grainstone, and the least abundant samples and the two barren ones came from lime mudstones. Nevertheless, the general reduction in abundance of conodonts is reflected in all lithologies in the Fillmore as compared with the same lithologies in the House.

The new data presented here show that the upper limit of the range of Fauna C does not coincide with the House–Fillmore boundary as reported previously (see Fig. 8.3). The relatively uncommon components of the faunal association were not found in the Fillmore, but the dominant members (*Acanthodus lineatus* Furnish, *Acanthodus uncinatus* Furnish, '*Paltodus' bassleri* Furnish, *Rossodus manitouensis* Repetski and Ethington, '*Oistodus' triangularis* Furnish, ?*Acanthodus uncinatus* Furnish, '*Pal-* which was collected 3.2 m (10.5 ft) above the base of the formation. Ethington and Clark (1982, Table 5) reported only a few elements from this part of the Fillmore at their Section C which is about a mile north of the locality studied here. The elements we found in the uppermost House and lowest Fillmore are very

small and slender in contrast to those from lower in the section, so that they may have passed through a sieve and been undetected in the earlier study.

The conodonts recovered above the basal 3 metres of the Fillmore are wholly different from those below that horizon. In addition to lesser abundance, the conodonts in the upper part of our measured section display lower diversity than those beneath. The most frequently occurring members of the population are a species of *Drepanoistodus*, an albid coniform species that we refer to *Oneotodus* Lindström, and *Scandodus* sp. 6 of Ethington and Clark (1982). None of these forms was considered a component of Fauna D by Ethington and Clark (1971), but they acknowledged that they had sparse recovery from the basal Fillmore. Subsequently they postulated the presence of an interval with a low-diversity fauna below the level of introduction of the conodonts they considered characteristic of Fauna D (*Glyptoconus quadraplicatus*, *Eucharodus parallelus*, *Acodus deltatus* Pander). Our results reinforce that interpretation. The present collections yielded a specimen that we tentatively identify with *E. parallelus* at 10.5 m (34.5 ft) in the Fillmore and another that is an S element of a species of *Acodus* close to *A. deltatus* at 11.7 m (38.5 ft). None of the species in this low-diversity fauna is an obvious descendant of a member of the more abundant and diverse Fauna C which they replace within an interval of no more than 0.6 m in the lower Fillmore.

### (d) Other fossils

Hintze (1951) subdivided the Ordovician rocks beneath the Eureka Quartzite (Middle Ordovician) in the Ibex area into a succession of faunal zones based primarily on the stratigraphical distribution of trilobites. Our measured section encompasses the upper part of his Zone B (*Symphysurina* Zone) and all of Zone C (*Paraplethopeltis* Zone), both within the upper House Limestone, as well as the

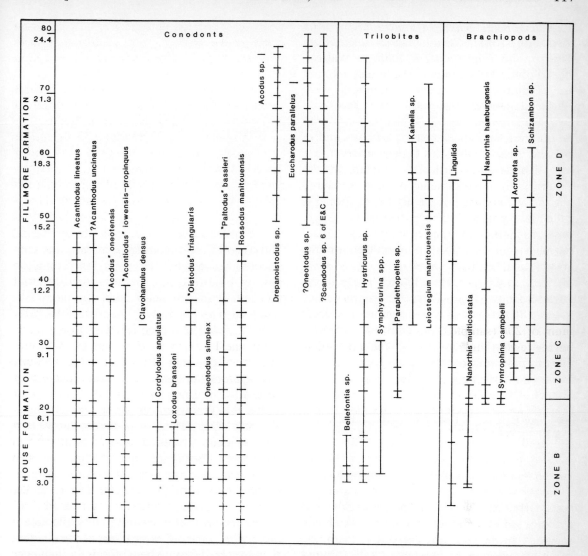

Fig. 8.3—Stratigraphical ranges of conodonts, trilobites, and brachiopods in the Ibex section; vertical scale is in feet (above) and metres (below). Trilobite zones of Ross (1951) and Hintze (1951) are shown in right column.

lower part of Zone D (*Leiostegium-Kainella* Zone) in the Fillmore. The entire sequence was examined carefully for trilobites, and all ledges in which they were seen were collected for these fossils. In addition, each of the samples obtained for conodonts was carefully split and examined before processing in order to add to the documentation of trilobite distribution within the measured section. As a conse-

quence, the ranges of the trilobites shown on Fig. 8.3 are comprehensive.

Trilobites recovered from the *Symphysurina* Zone include *Bellefontia chamberlaini* Clark, *Hystricurus genalatus* Ross, *Hystricurus paragenalatus* Ross, *Symphysurina globacapitella* Hintze, and *Clelandia utahensis* Ross. The overlying thin (3.28 m) *Paraplethopeltis* Zone contains *Paraplethopeltis genacurva* Hintze,

*Paraplethopeltis genarecta* Hintze, *Hystricurus genalatus* Ross, *Symphysurina uncaspicata* Hintze, and *Kainella* cf. *K. billingsi* Walcott. No trilobites were found in the lowest 4.4 m of the Fillmore, but modest numbers were obtained above this level. We recovered *Leiostegium manitouensis* Walcott, *Kainella* cf. *K. billingsi*, and species of *Hystricurus* and of *Parahystricurus* from this highest interval.

Articulate and inarticulate brachiopods are present but not abundant in our measured section. Abundance of the articulate species is much lower in the Fillmore than in the House, whereas inarticulate species are quite abundant through 6 m of the section centred on the formational contact. *Nanorthis multicostata*, Ulrich and Cooper has a limited range in the section from 3.7 to 7.6 m, and *Syntrophina campbelli* (Walcott) is even more restricted (from 6.7 to 7.3 m). In contrast *Nanorthis hamburgensis* (Walcott) and the inarticulate species have much longer ranges.

Fig. 8.4—Location of measured sections of upper McKenzie Hill and lower Cool Creek formations in southern Oklahoma.

## 8.3  ARBUCKLE GROUP, SOUTHERN OKLAHOMA

### (a) Section locations and lithostratigraphy

The Arbuckle Group in the Arbuckle and Wichita Mountains of southern Oklahoma includes strata that are correlative with the sampled interval in western Utah. Mound (1968) reported conodonts from a reconnaissance study of part of the Lower Ordovician in the Arbuckle Mountains. His paper lacks biostratigraphical documentation and presents an inconsistent taxonomic treatment of the conodonts. Nevertheless, the report clearly shows that conodonts of Fauna C occur at the top of the McKenzie Hill Formation and that the overlying Cool Creek Formation contains Fauna D. This sequence thus is an appropriate one for comparison of its conodonts and their stratigraphical ranges with those in the Ibex area.

We collected samples across the McKenzie Hill–Cool Creek boundary at two sections, selected because of accessibility and completeness of exposure and because their trilobite faunas had already been studied in detail (see Fig. 8.4). The Highway 77 Section (Stitt 1983; NW1/4, Sec. 12, T2S, R1E, Turner Falls, Oklahoma 7 1/2 min. Quadrangle) is located on the Chapman Ranch in the western Arbuckle Mountains. The Chandler Creek Section (Stitt 1983; NW1/4, NW1/4, Sec. 14, T4N, R12W, Richards Spur, Oklahoma 7 1/2 min. Quandrangle) is in the foothills of the Wichita Mountains of southwestern Oklahoma. Both sections are on privately owned ranches, and permission for study must be obtained from the owners. The rocks dip steeply, and the slopes are gentle at both exposures, so that measuring and collecting is easy. The variable responses to weathering displayed by different beds within the sequence cause some of them to project prominently as continuous exposures that can be followed readily across the countryside, whereas other beds underlie grassy slopes. Short offsets from the line of section allowed sampling of the recessive units, and an essentially complete succession of strata was studied.

Each of the two sections was sampled through an interval of 18.3 m centred on the formational contact. In the absence of a profound change in lithology at the contact, we followed Ham (1969) and Stitt (1983) who defined the base of the Cool Creek at the level of abrupt introduction of abundant quartz sand, siliceous oolites, and stromatolites in the limestones of the sections. Thirty-one samples were taken at each locality, beginning at the formational contact and continuing at 0.3 m (1 ft) intervals through the next 3.1 m (10 ft) above and below it. Additional samples were taken at 3.7, 4.6, 6.1, 7.6, and 9.1 m above and below. The primary lithologies encountered within the sampled succession are lime mudstones and peloidal grainstones. Stitt (1983) postulated that the increasing abundance of lime grainstones relative to lime mudstones with increasing height in the McKenzie Hill is an indication of progressive shallowing over time so that ever greater proportions of mud were removed by winnowing from the accumulating sediment. Evidence for continued shallowing during deposition of the lower part of the Cool Creek Formation is provided by the occurrence of digitate stromatolites and oolites.

A clearly defined discontinuity surface was recognized during petrographic study of a hand specimen collected between 2.1 and 2.4 m above the base of the Cool Creek Formation in the Chandler Creek outcrop, and this subtle surface subsequently was found in the outcrop. This hardground is just above the highest occurrence of Fauna C and slightly below the lowest level (2.74 m in the Cool Creek) at which a younger conodont fauna is introduced. This relationship suggests that abrupt faunal replacement in this section is the result of unconformity. Careful scrutiny failed to detect a comparable discontinuity separating the same two conodont populations at the Highway 77 Section, so the presence of the hardground between the two faunas at Chandler Creek is probably fortuitous.

### (b) Conodont biostratigraphy

A kilogram of rock from each of the samples was processed for recovery of conodonts. Five samples from the Chandler Creek Section were barren; the others from this locality produced

1004 specimens. In contast, the Highway 77 Section had 10 barren samples and produced only 537 specimens from the productive ones. The different frequencies of occurrence in the two correlative sections may be the result of their respective paleoenvironmental settings. Stitt (1983, p. 5–6) suggested that at the time of deposition of the Lower Ordovician rocks the Chandler Creek area was deeper and/or quieter than the Arbuckle Mountains area owing to their positions relative to the axis of the Southern Oklahoma Aulacogen.

Conodonts are less abundant in the basal Cool Creek ($\bar{x}$ = 19 elements/sample at Chandler Creek, 9.5 elements/sample at Highway 77) than in the upper beds of the McKenzie Hill ($\bar{x}$ = 57.5 elements/sample at Chandler Creek, 47.7 elements per sample at Highway 77). As is the case in the Ibex area, the reduction in frequency occurs at the formational boundary. Conodonts of Fauna C are present throughout the sampled part of the McKenzie Hill at the Chandler Creek Section (Fig. 8.5) and some members of this association continue through the lower 2.1 m of the overlying Cool Creek. In contrast, Fauna C is restricted to the McKenzie Hill at Highway 77, as reported by Mound (1968). This seeming discrepancy between the lithological sequence and the stratigraphical distribution of the conodonts at the two sections may reflect the way in which the formational boundary is identified. The influx of quartz sand into the basin that is used to define the base of the Cool Creek probably did not reach the two sections, which are approximately 120 km apart, at the same time, and the formational boundary must be diachronous.

The occurrences of Fauna C in the Arbuckle and Wichita Mountains are dominated in abundance and in stratigraphical distribution by the same species that are dominant at Ibex. As is the case in western Utah, *Loxodus bransoni*, '*Acodus*' *oneotensis*, and *Cordylodus angulatus* are present but rare, and they occur sporadically among the samples. Species that are present in Oklahoma but only rarely represented or absent in the Utah section are *Clavohamulus*

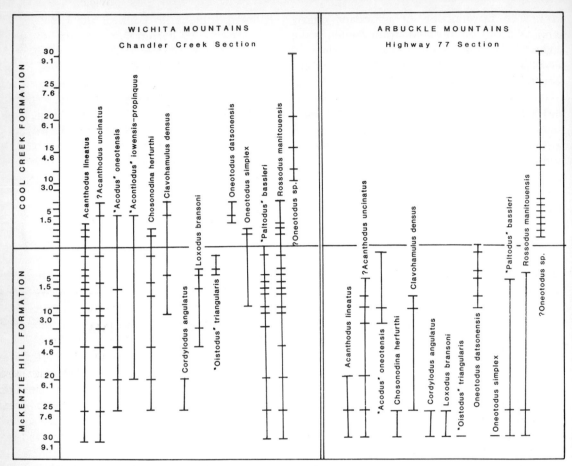

Fig. 8.5—Distribution of conodonts in the Oklahoma sections; vertical scale is in feet (above) and metres (below).

*densus* Furnish, *Chosonodina herfurthi* Müller, *Oneotodus datsonensis* Druce and Jones, and a species of ?*Semiacontiodus* Miller. With the exception of these forms, the composition of Fauna C in southern Oklahoma is much the same, quantitatively and qualitatively, as it is in western Utah.

At both Oklahoma sections the rocks above those with Fauna C yielded an impoverished fauna of very low diversity. This sparse population consists wholly of albid, coniform elements of varying cross-sections that we are assigning tentatively as a species of *Oneotodus* Lindström. No species that are typical of Fauna D were found in either of our measured sections, but Mound (1968) reported some of

them from the basal part of the Cool Creek at Highway 77. Because his identifications of conodont taxa were not consistent, these low stratigraphical occurrences of conodonts of Fauna D are suspect. Nevertheless, such species as *E. parallelus*, *G. quadraplicatus*, *Ulrichodina* spp., and *Macerodus dianae* are represented among the conodonts that Mound illustrated from the Highway 77 Section. In another collection from that section, made by R. L. Ethington, the lowest occurrence of *G. quadraplicatus* is at 202 m above the base of the Cool Creek, and the typical components of Fauna D become increasingly abundant and dominant in samples from higher in the formation.

## (c) Other fossils

The Mckenzie Hill Formation at the sections considered here has been studied carefully by Stitt (1983) who recorded the presence of trilobites at both localities. He recognized assemblages of species in the Chandler Creek Section that he compared with those of the *Symphysurina* and *Paraplethopeltis* zones of western Utah, but only the former occurs at Highway 77. The uppermost 26.8 m at Highway 77 as well as the lower Cool Creek at these localities have not been productive of trilobites in Stitt's studies. We did not find identifiable specimens in the more restricted stratigraphical interval that we examined, although cross-sections of carapaces were observed in thin sections. At present, conodonts are the only fossils known from both the Utah and Oklahoma sections.

## 8.4  DISCUSSION

The palaeogeographical and sedimentological setting of central and eastern United States during Early Ordovician times was summarized by Ross (1976). According to his interpretation, the Oklahoma sections were located at about 20°S latitude when the McKenzie Hill and Cool Creek formations were being deposited, at which time the Ibex area was situated to the northwest of them and closer to the equator. Both regions were in the wind-belt of the southern Tropical Easterlies. At that time the Sauk Sea was at or near its maximum extent, and the North American craton was awash except for the Canadian Shield. The broad but shallow Sauk Sea was interrupted by the Transcontinental Arch, an axis of shallow sea-floor and islands that extended from the present state of Arizona to Minnesota. Ordovician rocks are generally absent from the region of the Transcontinental Arch, and stratigraphical units commonly thin towards it. Clearly, the Arch restricted circulation between the ocean-facing shelf then located to the north of it and

the cratonal platform to the south where the Oklahoma sections are located (see Fig. 8.6).

Open ocean conditions prevailed to the north of the Transcontinental Arch during Early Ordovician time, and a broad, thick succession of marine limestones was deposited. The sedimentological setting within which the House Limestone accumulated has been interpreted by Cook and Taylor (1977) as a shelf that was prograding away from the Arch, so that the House represents progressively shallower-water deposition with height above its base. This shallowing trend continued during deposition of the Fillmore as indicated by the increasing abundance of flat-pebble conglomerates as compared with the upper House and by the presence of 'algal reefs' (Hintze 1973) in the lower 60 m of the Fillmore.

The Lower Ordovician rocks across the broad central platform of the United States consist largely of dolostones whose low-diversity faunas and sedimentary structures (e.g. oolites, mud cracks, domal stromatolites) argue for deposition in very shallow and perhaps, at times, hypersaline waters. The Lower Ordovician Arbuckle Group of southern Oklahoma is an exception to this depositional pattern. It has yielded a diverse fauna unlike that in coeval rocks in contiguous areas, and was deposited under normal marine conditions. The Lower Palaeozoic setting in this region of the southern midcontinent has been interpreted as an aulacogen (Hoffman, Dewey, and Burke 1974), an aborted continental rift that opened southeasterly (modern orientation). Subsidence in this basin was rapid, but deposition kept pace with it, for nearly 3000 m of relatively shallow-water limestones were deposited in this trough during Early Ordovician time. The frequency and abundance of oolitic, sandy, and stromatolitic beds in the succession and particularly in the Cool Creek Formation suggest that the depth of water generally was less than in the Ibex area of Utah during deposition of the House–Fillmore sequence. The depth of water in the Southern Oklahoma Aulacogen probably fluctuated

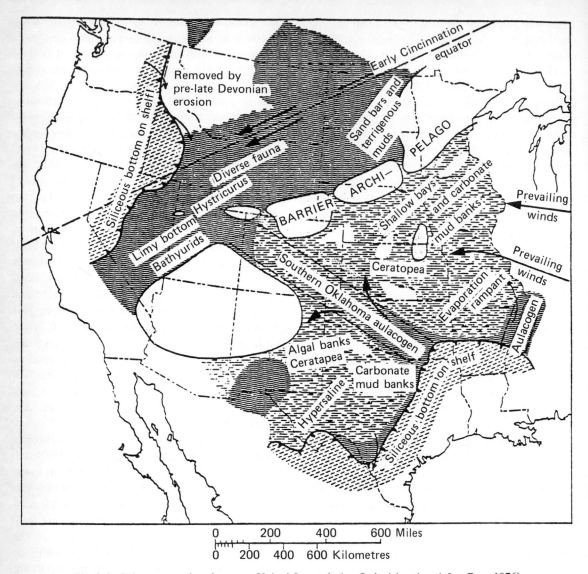

Fig. 8.6—Palaeogeography of western United States during Ordovician time (after Ross 1976).

repeatedly during the Early Ordovician in response to changes in intensity of subsidence or to its periodicity, and to variations in the rate at which calcareous sediment was supplied. As reported by Stitt (1983), the part of the section that is under consideration here was deposited during a period of shallowing.

The same general succession of trilobites is known in the House Limestone and in the lower McKenzie Hill Formation, although

these fossils have not been found in the part of the latter unit that we sampled. Stitt (1983) reported the trilobite succession in Oklahoma to include a lower zone of *Symphysurina*, an intermediate *Bellefontia-Xenostegium* Zone, and an upper *Paraplethopeltis* Zone, and he was able to correlate this succesion, which includes all but the upper part of the McKenzie Hill, with Hintze's trilobite zones at Ibex.

Despite their geographical separation of

1700 km and contrasting geographical settings on opposite sides of the Transcontinental Arch, the above palaeogeographical and biostratigraphical comparisons indicate that the Lower Ordovician rocks of southern Oklahoma and western Utah are sufficiently similar to permit comparisons of their respective conodont faunas without major environmental concerns. As reported above, the basic pattern of conodont distribution is the same in both areas. The same species (*A. lineatus, R. manitouensis*) are dominant in Fauna C at all three of the sections that we sampled, and the species of lesser abundance (e.g. *L. bransoni, C. angulatus*) are the same as well. The only significant difference is the presence of *Clavohamulus densus* and *Chosonodina herfurthi* in Oklahoma. These species were not recovered at Ibex in this study, and were found to be rare in the more comprehensive earlier study by Ethington and Clark (1982). Their greater abundance in Oklahoma may be indicative of the shallower-water conditions that we believe to have prevailed there. Ethington and Repetski (1984) interpreted *C. densus* as a stenotopic form based on the known geographical and stratigraphical distributions of the taxon, but they did not identify the conditions that regulated its distribution. The same authors inferred a deep-water habitat for *C. herfurthi*; they were influenced in this conclusion by the presence of this species in continental-slope deposits in the Ouachita Mountains of west-central Arkansas. The much greater abundance of *C. herfurthi* in the McKenzie Hill is not consistent with a deep-water habitat, and the scarcity of its occurrence in the House Limestone suggests that the species did not flourish in the depths of water of open shelves.

The rocks that occur just above the range of Fauna C contain sparse conodont faunas of low diversity in all three of our sampled sections. The dominant components of the population in both Oklahoma and Utah are largely albid, coniform species that we identify with *Oneotodus* Lindström. These forms represent the entire population in our samples from Oklahoma, but they are associated with *Scandodus* sp. 6 of Ethington and Clark (1982) in Utah. In both regions a more diverse population that includes characteristic and long-ranging species of Fauna D (*G. quadraplicatus, E. parallelus, S. rex, Ulrichodina* spp., *Macerodus dianae*) is introduced higher in the section. The disappearance of Fauna C and the introduction of the low-diversity fauna that replaces it occur within a very short stratigraphical interval, and no mixing of the two populations is indicated for any of the studied sections. The faunal replacement in each of the sections occurs within a stratigraphical sequence that displays evidence for gradually shallowing depth of deposition. No abrupt lithological changes are present. The hardground observed in the Chandler Creek Section just above the highest occurrence of Fauna C is not duplicated at the other two sections, so we conclude that its position relative to the level of faunal replacement is not significant. We are unable to identify a physical attribute that is common to the three sections that we can correlate with the faunal change.

Comparable general successions of conodonts are known from many places in North America and from the Siberian Platform; these occurrences were summarized by Ethington and Clark (1982, p. 608). Owing to lack of continuous exposures or to broadly spaced sampling, stratigraphical documentation of the ranges of conodonts is lacking for all of these sections. All of these other occurrences are consistent with the generalized biostratigraphical succession of conodonts outlined by Ethington and Clark (1971), but the presence of a low-diversity fauna between the range limits of Faunas C and D is not demonstrated. Recent work in central Missouri (Chapman 1984) shows Fauna C to be present in the upper Gasconade Formation, the impoverished fauna to be present but very rare in the overlying Roubidoux Formation, and typical conodonts of Fauna D to occur in the still younger Jefferson City Formation. Recoveries of conodonts in that study were low, and the sequence from which they were obtained is a composite section based on several localities. These conodonts, however, support the contention that the relations reported here for the House Limestone and Fillmore Formation in Utah and for the lower Arbuckle Group in Oklahoma also obtain within the widespread shallow-water dolostones of the North American Platform.

## 8.5   COMPARISON WITH TRILOBITE BIOMERES

Palmer (1965) recognized regional biostratigraphical units that he termed 'biomeres' within the Upper Cambrian of the western United States on the basis of non-evolutionary changes in the dominant elements of the trilobite succession. The biomere concept has been widely applied in the biostratigraphy of Cambrian rocks, and Stitt (1983) has used it to interpret the distribution of some Lower Ordovician trilobites. Biomeres are characterized by the abrupt upward termination of an assemblage of species that dominated the trilobite population during the deposition of a significant stratigraphical succession. In higher beds, and commonly no more than a few centimetres higher, a low-diversity fauna is introduced; this fauna consists of species that represent different evolutionary lineages from those of the population that they replaced. The new fauna underwent evolutionary diversification, and then it, in turn, experienced abrupt disappearance.

Biomere boundaries, and the faunal replacements by which they are defined, have been interpreted as the result of responses to major environmental changes that were activated over broad areas of the sea floor during short periods. The cause of extinction is not known, although Stitt (1977) suggested that episodes of global cooling brought broad areas of shallow sea floor below the thermocline. Under these conditions, trilobites adapted to life in warm but shallow seas became extinct and were replaced by immigrant species that

formerly had been relegated to deeper and cold environments. Such events are assumed to have been of brief duration, so that biomere boundaries, although considered to be somewhat diachronous, have been treated as essentially chronostratigraphical in nature. A dissenting opinion has been offered by Ludvigsen and Westrop (1983) who argue that the faunal changes at biomere boundaries are accompanied by changes in lithofacies, that these changes do not reflect nearly catastrophic events, and that they are not reliable for correlation over long distances.

The stratigraphical distributions of conodonts in our measured sections in Utah and Oklahoma show marked similarities to the patterns displayed by trilobites from sequences of rocks that cross biomere boundaries. In each case, a well-established population of conodonts disappears abruptly from the section without an accompanying change in lithology of the enclosing rocks. This phenomenon corresponds to the fourth and youngest of the developmental stages that Stitt (1971) recognized as typical of a trilobite-based biomere. A sparse fauna of low diversity is introduced in each section less than a metre above the highest occurrence of the older conodont fauna. This impoverished assemblage is comparable with the trilobites found in Stage 1 of Stitt's succession. At still higher stratigraphical levels and above the tops of the intervals here studied in detail, a more diversified population is established and is maintained through a significant stratigraphical sequence. Stitt considered the introduction of such long-ranging and morphologically stable forms to be characteristic of Stage 2 of a typical biomere. Obvious ancestors of the younger conodont faunas cannot be identified among the members of the oldest population, so that the younger forms must have migrated into the studied areas from elsewhere.

Although close biostratigraphical control is not available, similar sequences of conodonts are known from numerous localities in North America and Siberia, so that this faunal extinction–replacement event appears to be represented widely in the Midcontinent Province. The event can be identified not only in normal marine facies such as those discussed here, but also in the shallow-water dolostones of central United States (Furnish 1938, Chapman 1984). It thus seems to offer the possibilities for high-resolution correlation that Ethington and Clark (1982) suggested.

The reason for this significant development in the history of Lower Ordovician conodonts is not clear. Disagreement exists regarding the life-style of conodonts (Seddon and Sweet 1971, Barnes and Fåhræus 1975, Klapper and Barrick 1978). In their review of the palaeogeographical distribution of Lower Ordovician conodonts in North America, Ethington and Repetski (1984) concluded that known occurrences support recognition of shallow-water near-shore populations that contrast with associations that lived in waters of intermediate and basinal depths. They did not make any inferences as to whether the organisms occupied a benthic habitat or lived above the substrate. Further, they were unable to identify specific environmental factors among the many that vary along a bathymetric profile (pressure, temperature, salinity, turbidity, etc.) as regulators of the distribution of the conodonts.

Lithological evidence indicates that the faunal replacement recorded here took place during a time of progressive shallowing of the sea at each of the three sections that we studied. Evidence for this shallowing is present in rocks beneath the lowest horizons that we sampled, and the shallowing continued after the deposition of the youngest rocks that we sampled. This shallowing may be leading to the global sea-level minimum that Vail *et al*. (1977, fig. 1) recognized at the end of the Early Ordovician. Changing conditions at the substrate and in the water column during this interval of shoaling must have been responsible for the faunal replacement discussed here and may be reflected in the stratigraphical distributions of other fossils from the same sections.

Stitt (1983) recognized a Symphysurinid Biomere from the distribution of trilobites in the Utah and Oklahoma sections that we studied. He reported, following Ethington and Clark (1982), that the conodonts of Fauna C disappear from the Utah section at the same stratigraphical level as the trilobites of this biomere, and concluded that the same event was responsible for the extinction of both groups. The present study dictates revision of that conclusion. Conodonts and trilobites occur together throughout the entire sequence that we sampled at the Ibex locality. There, the extinction of the conodonts of Fauna C occurs 4.27 m higher than that of the trilobites of the Symphysurinid Biomere (top of Zone C of Hintze 1951), so that a simple catastrophic event, such as impact of an object from an extraterrestrial source, was not responsible. It also may be significant that the other benthic group recovered at Ibex, the brachiopods, was not influenced as were the trilobites and the conodonts. Long-ranging forms continue through the sampled sequence without interruption. We conclude that the independent responses of the conodonts and trilobites are to dynamic environmental changes that accompanied the shallowing of the sea over broad areas of North America. The two groups did not respond to the same factors, or did not respond at the same level of development of those factors, so that the conodont extinction occurred somewhat later than that of the trilobites.

# ACKNOWLEDGEMENTS

Engel's work in Utah was supported by a grant from the Geological Society of America and by the E. J. Palmer Award from the University of Missouri-Columbia. The M. G. Mehl Award from UMC provided field expenses for Elliott's work in Oklahoma. Landowners' permission to collect the Oklahoma sections is appreciated. We are indebted to Louis M. Ross who prepared the SEM photographs of specimens illustrated in Plate 8.1. The SEM facility at UMC was established in part by grant EAR-8217931 from the National Science Foundation.

## REFERENCES

Abaimova, G. P. 1975. Early Ordovician conodonts of the middle fork of the Lena River. *Trudy Nauchno-Issledovatelskogo Instituta, Geologii, Geofiziki i Mineralnogo Sirya* **207** 129 pp., 10 pls.

Barnes, C. R. and Fåhræus, L. E. 1975. Provinces, communities, and the nektobenthic habit of Ordovician conodontophorids. *Lethaia* **8** 133–149.

Chapman, K. R. 1984. Conodonts from the Lower Ordovician of central Missouri. *Unpublished M.S. thesis*, University of Missouri-Columbia, Columbia, Missouri, 73 pp., 1 pl.

Cook, H. E. and Taylor, M. E. 1977. Comparison of continental slope and shelf environments in the Upper Cambrian and Lower Ordovician of Nevada. In: H. E. Cook and P. Enos (eds.): *Deep Water Carbonate Environments, Society of Economic Paleontologists and Mineralogists Special Publication* **25** 51–82.

Ethington, R. L. and Clark, D. L. 1971. Lower Ordovician conodonts in North America. In: W. C. Sweet and S. M. Bergström (eds.): *Symposium on Conodont Biostratigraphy, Geological Society of America Memoir* **127** 63–82, 2 pls.

Ethington, R. L. and Clark, D. L. 1982. Lower and Middle Ordovician conodonts from the Ibex Area, western Millard County, Utah. *Brigham Young University Geology Studies* **28** part 2, 155 pp., 14 pls.

Ethington, R. L. and Repetski, J. E. 1984. Paleobiogeographic distribution of Early Ordovician conodonts in central and western United States. In: D. L. Clark (ed.): *Conodont Biofacies and Provincialism, Geological Society of America Special Paper* **196** 89–102.

Furnish, W. M. 1938. Conodonts from the Prairie du Chien beds of the Upper Mississippi Valley. *Journal of Paleontology* **12** 318–340, pls. 41, 42.

Ham, W. E. 1969. *Regional Geology of the Arbuckle Mountains, Oklahoma*. Oklahoma Geological Survey Guide Book 17, 52 pp.

Hintze, L. F. 1951. Lower Ordovician detailed stratigraphic sections for Western Utah. *Utah Geological and Mineralogical Survey Bulletin* **39**, 99 pp.

Hintze, L. F. 1973. Lower and Middle Ordovician stratigraphic sections in the Ibex area, Millard County, Utah. *Brigham Young University Geology Studies* **20** part 4, pp. 3–36.

Hoffman, P., Dewey, J. F., and Burke, K. 1974. Aulacogens and their genetic relation to geosynclines, with a Proterozoic example from Great Slave Lake, Canada. In: R. H. Dott, Jr and R. H. Shaver (eds.): *Modern and Ancient Geosynclinal Sedimentation. Society of Economic Paleontologists and Mineralogists Special Publication* **19** 38–55.

Klapper, G. and Barrick, J. E. 1978. Conodont ecology; pelagic versus benthic. *Lethaia* **11** 15–23.

Lindström, M. 1955. Conodonts from the lowermost Ordovician strata of south-central Sweden. *Geologiska Föreningens i Stockholm Förhandlingar* **76** 517–604, pls. 2–7.

Ludvigsen, R. and Westrop, S. R. 1983. Trilobite biofacies of the Cambrian-Ordovician boundary interval in northern North America. *Alcheringa* **7** 301–319.

Moskalenko, T. A. 1967. Conodonts of the Chunsk Stage (Lower Ordovician), River Moiero and Podkamennaya Tunguska. In: *New Data on the Biostratigraphy of the Lower Paleozoics of the Siberian Platform*, Izdatelstvo 'Nauka', 98–116, pls. 22–25.

Mound, M. R. 1968. Conodonts and biostratigraphy of the lower Arbuckle Group (Ordovician), Arbuckle Mountains, Oklahoma. *Micropaleontology* **14** 393–434, 6 pls.

Nowlan, G. S. 1985. Late Cambrian and Early Ordovician conodonts from the Franklinian miogeosyncline, Canadian Arctic Islands. *Journal of Paleontology* **59** 96–122.

Palmer, A. R. 1965. Biomere – A new kind of biostratigraphic unit. *Journal of Paleontology* **35** 149–152.

Repetski, J. E. 1982. Conodonts from El Paso Group (Lower Ordovician) of Westernmost Texas and Southern New Mexico. *New Mexico Bureau of Mines and Mineral Resources Memoir* **40** 121 pp., 2 pls.

Ross, R. J. Jr. 1951. Stratigraphy of the Garden City Formation in Northeastern Utah and its Trilobite Faunas. *Yale University Peabody Museum of Natural History Bulletin* **6** 161 pp., 36 pls.

Ross, R. J. Jr 1976. Ordovician sedimentation in the western United States. In: M. G. Bassett (ed.): *The Ordovician System: Proceedings of a Palaeontological Association Symposium, Birmingham, September, 1974*. University of Wales Press and National Museum of Wales, Cardiff, 73–105.

Sando, W. J. 1958. Lower Ordovician section near Chambersburg, Pennsylvania. *Geological Society of America Bulletin* **69** 837–854, 2 pls.

Seddon, G. and Sweet, W. C. 1971. An ecologic model for conodonts. *Journal of Paleontology* **45** 869–880.

Sergeeva, S. P. 1966. Biostratigraphic ranges of conodonts in the Tremadocian Series (Ordovician) of the Leningrad District. *Doklady Akademii Nauk SSSR* **167** 672–674.

Stitt, J. H. 1971. Repeating evolutionary pattern in Late Cambrian trilobite biomere. *Journal of Paleontology* **45** 178–181.

Stitt, J. H. 1977. Late Cambrian and Earliest Ordovician Trilobites, Wichita Mountains Area, Oklahoma. *Oklahoma Geological Survey Bulletin* **124** 79 pp., 6 pls.

Stitt, J. H. 1983. Trilobites, Biostratigraphy, and Lithostratigraphy of the McKenzie Hill Limestone (Lower Ordovician), Wichita and Arbuckle Mountains, Oklahoma. *Oklahoma Geological Survey Bulletin* **134** 54 pp., 6 pls.

Vail, P. R., Mitchum, R. M., Jr., and Thompson, S. III, 1977. Seismic stratigraphy and global changes of sea level, Part 4: Global cycles of relative changes of sea level. In: C. E. Payton (ed.): *Seismic Stratigraphy – Applications to Hydrocarbon Exploration. American Association of Petroleum Geologists Memoir* **26** 83–97.

Viira, V. 1974. Ordovician Conodonts of the East Baltic. *Institut Geologii Akademii Nauk Estonskoi SSR*, 142 pp., 13 pls.

# 9

# Lithological and conodont distributional evidence for episodes of anomalous oceanic conditions during the Silurian

L. Jeppsson

## ABSTRACT

The early Wenlock extinction event at the end of the *Pterospathodus amorphognathoides* Zone was preceded by widespread deposition of atypical sediments. All platform-equipped conodonts, most of those with ramiform elements and several of those with coniform elements (*Panderodus langkawiensis, P. recurvatus, Pseudooneotodus tricornis* and gen. et sp. nov.) disappeared from studied sedimentary sequences. However, some apparently escaped extinction. It seems that these persisted, confined to unknown refuges, to reappear briefly worldwide during another episode of atypical sedimentation in the Ludlow, at the time of development of the *Polygnathoides siluricus* Zone and/or during the preceding *Ancoradella ploeckensis* Zone. Similar episodes on a more minor scale have also been recognised.

## 9.1 INTRODUCTION

Of the diverse organisms that occurred in the marine realm during the Silurian Period, conodonts were particularly widespread and abundant and are potentially important in studies of episodic, rapid changes in oceanic conditions. Taxonomic and biostratigraphical studies have now reached a level where some of this potential can be realized.

The early Wenlock extinction event, which severely affected conodonts (Aldridge 1976, Aldridge and Jeppsson 1984), has not previously been included in discussions of extinctions and contemporaneous lithological and other abiotic changes (Alvarez *et al*. 1980, McLaren 1983, House, 1985). However, the extinction in the earliest Wenlock seems to have coincided with a change from more oxygenated oceanic conditions to 'greenhouse' conditions, with a final fauna before the extinction that thrived during an episode of widespread abnormal sedimentary conditions. Similar changes seem to have accompanied other Silurian events.

Discussions of extinctions and other faunal changes must be based on terminations of lineages and not on nomenclatural boundaries. Thus, in any analysis, the species concept used should be explained; for the Silurian conodonts considered here a species embraces the whole known part of a lineage. It is thus a much wider concept than that used for many other taxa.

Each species exhibits a considerable temporal and geographical variation, and the concept approaches that used for Recent birds and larger mammals. Similarly, most genera include several lineages, so that at each point in time they were represented by several species, as in most Recent taxa.

## 9.2 POST-LLANDOVERY SILURIAN FAUNAS WITH PLATFORM-EQUIPPED TAXA

Two of the most characteristic and most widely identified Silurian conodont zones are the *Pterospathodus amorphognathoides* Zone (*P. am.* Z.) and the *Polygnathoides siluricus* Zone (*P. s.* Z.). There have been different definitions of the boundaries of the *P. am.* Z., but here I follow Walliser's (1964) definition (see also Barrick 1983)—the base coincides with the appearance of the eponymous taxon and the top with its disappearance. Reference herein to the *P. s.* Z. relates only to the time of widespread presence of *P. siluricus*. As discussed by Thorsteinsson and Uyeno (1981) and by Jeppsson (1983) there are a few records of *P. siluricus* together with graptolites of the *S. chimaera* Zone, but in most areas where conodonts of that age are known the only platform-equipped taxa recorded are *K. variabilis*, *A. ploeckensis*, and '*Spathognathodus inclinatus*' *hamatus*. For the record, the last two have now been found on Gotland: *A. ploeckensis* at Snoder 3, sample G71-10LJ, and Rangsarve 1, sample G75-57CB; '*S. i.*' *hamatus* at Gardsby 1, sample G82-27LJ.

*Pterospathodus amorphognathoides* has a very wide distribution, both ecologically (Aldridge and Mabillard 1981, Mabillard and Aldridge 1983) and geographically. A large number of species have been recorded from the *P. am.* Z., and these are listed on Table 9.1. In many areas the *P. am.* Z. is condensed, preventing a detailed study of the transition to succeeding strata. One exception is on Gotland, where the exposed strata include some metres of Llandovery marls (Jeppsson 1983) and a complete sequence of Wenlock and Ludlow strata. The total thickness of Silurian sediments exposed is about 500 m (Hede 1960). The *P. am.* Z. is found in the lowest unit, the Lower Visby Beds, and is exposed to a thickness of at least 11 m. Faunal changes through this zone into the succeeding strata have been investigated in detail at the section at Ireviken 3 (for locality information see Laufeld 1974, Odin *et al.* 1984), where the lower of two prominent bentonite layers, 2.46 m apart, serves as a reference level. A change in clay mineralogy, weathering style, and fauna occurs at the base of the marls of the Upper Visby Beds, about 5.4 m above the reference level. The lowest strata exposed in the section contain 400 or more conodont elements per kg, but the frequency gradually decreases to about 50–300 in the lowest Upper Visby Beds, where more than 75% of the specimens belong to *Panderodus*. The highest specimens of *P. amorphognathoides* have been found at 3.96 m above the reference level (Table 9.1), although unidentifiable *Pterospathodus* juveniles occur in succeeding samples through to the lower Upper Visby Beds.

---

Table 9.1—Conodont taxa known from the *P. amorphognathoides* Zone, based mainly on the regional summary by Barrick (1983), various publications by Aldridge and his students (Aldridge 1972, 1974, 1975, 1976, Aldridge and Mabillard 1981, Aldridge and Mohamed 1982, Mabillard and Aldridge 1983, 1985), the paper by Savage (1985), and my own data from Gotland. My approach has been conservative, in order not to overestimate the extinction at the end of the zone.

The last known occurrence is given as follows: (1) probably close to the lower boundary, (2) probably at or close to the upper boundary, (3) interrupted occurrence, reappearing at the time of Slite f and/or in the mid-Ludlow, (4) apparently unaffected except for changes in frequency. The right-hand column gives the distance above the reference level of the appearance or disappearance of those taxa recognised at Ireviken 3. G = taxa at Ireviken unaffected by the extinction, (G) = taxa found in the *P. am.* Z. elsewhere on Gotland. The coral *Paleocyclus porpita* ranges up to 5.46 m.

| PLATFORM-EQUIPPED SPECIES | | |
|---|---|---|
| Apsidognathus ruginosus  Mabillard & Aldridge | 2 | |
|          tuberculatus  Walliser | 2 | |
|          walmsleyi  Aldridge | 2 | to 2·37m |
| Tuxekania barbarajeannae  Savage | 2 | |
| Johnognathus huddlei  Mashkova | 2 | to 5·32 m |
| Distomodus nodosus  Helfrich | 2 | |
|         staurognathoides  (Walliser) | 2 | to 5·96m |
| Aulacognathus bullatus  (Nicoll & Rexroad) | 1 | (G?) |
|         kuehni  Mostler | 1 | |
|         latus  (Nicoll & Rexroad) | 1 | |
| Neospathognathodus chapini  Savage | 2 | |
| Icriodella inconstans  Aldridge | 2 | |
| Icriodella ? sandersi  Mabillard & Aldridge | 2 | |
| Astropentagnathus irregularis  Mostler | 1 | |
| Pterospathodus amorphognathoides  Walliser | 2 | to 3·96 m |
|         pennatus  (Walliser) | 2 | to 5·76 m |
| Kockelella ranuliformis  (Walliser) | 2 | from 1·55m |

| SPECIES WITH BLADE AND RAMIFORM OR ONLY RAMIFORM ELEMENTS | | |
|---|---|---|
| Pterospathodus celloni  (Walliser) | 1 | |
| Carniodus carnulus  Walliser | 2 | to 3·96 m |
| 'Rotundacodina' aff. 'R.' dubia ( Rhodes ) of Mabillard & Aldridge | 2 or 3 | |
| Ozarkodina excavata ( Branson & Mehl) | 4 | G |
|         gulletensis  (Aldridge) | 2(4?) | |
|         hadra (Nicoll & Rexroad) | 2 | |
|         polinclinata ( Nicoll & Rexroad) | 2 | to 2·37 m |
|         sp. a | 2 | to 4·61 m |
| Aspelundia capensis  Savage | 2 | |
| Ligonodina cf. L. kentuckyensis  Branson & Branson | 2(4?) | G |
|         fluegeli  (Walliser) sensu Aldridge, 1979 | 2 | |
|         petila  Nicoll & Rexroad | 2 | to 4·61 m |
|         sp. b. (= Savage, 1985, fig. 10 F-L, O-S ) | 2 | to 2·37m |
| Tokeenia furcata  Savage | 2 | |

| SPECIES WITH ONLY CONIFORM ELEMENTS | | |
|---|---|---|
| Decoriconus fragilis ( Branson & Mehl) | 4 | G |
| Dapsilodus obliquicostatus ( Branson & Mehl) | 4 | (G) |
|         praecipuus  Barrick | 4 | |
|         sparsus  Barrick | 4 | |
| Walliserodus curvatus (Branson & Branson ) | 4 | |
|         sancticlairi  Cooper | 4 | |
|         sp. | 4 | to 5·56 m |
| Belodella n. sp. a. of Miller, 1978 | 2 | |
| Panderodus equicostatus  (Rhodes) | 4 | from 2·47m |
|         unicostatus  ( Branson & Mehl) | 3 or 4 | G |
|         sp. b | 2 ? | (G) |
|         sp. g. of Jeppsson, 1983 | 3 | to 2·16 m |
|         langkawiensis (Igo & Koike) | 2 | to 4·61m |
|         recurvatus (Rhodes ) | 3 | to 5·32 m |
| Pseudooneotodus beckmanni bicornis  Drygant | 4 | from 1·1 m |
|         tricornis  Drygant | 2 | to 0·85m |
| Gen. et sp. nov. | 2 | to 0·85m |

The distribution of taxa in the Ireviken 3 section is indicated on Table 9.1. The first faunal changes occur between 0.90 and 1.10 m above the reference level, where *Pseudooneotodus tricornis* is replaced by *Ps. beckmanni bicornis* and gen. et sp. nov. disappears. *Kockelella ranuliformis* is rare, but has not been found at all in association with *Ps. tricornis*. A sample from 2.36–2.46 m contains the last specimens of *Ozarkodina polinclinata*, *Ligonodina* sp. b. and *Apsidognathus walmsleyi*. *Panderodus equicostatus* appears to replace *Panderodus* sp. g in this sample, too, but it is known from much older strata in the När core (see Snäll 1977, for locality), and this is a level of reappearance rather than first appearance. The highest *Panderodus langkawiensis*, *Ligonodina petila* and *Ozarkodina* sp. a are found in a large sample from 4.61 m, but their frequency is very low in this part of the section and they may well range higher.

*Pterospathodus*, *Walliserodus*, *Johnognathus*, and *Distomodus* range into the lower part of the Upper Visby Beds. The dark colour of several specimens suggests that they are reworked, but some are of normal colour and may be indigenous. In all cases, it is to be expected that bioturbation will have extended the observed ranges somewhat beyond the original biological ones. However, the evidence indicates that there is a small but real spread in the time of the local extinction. If the rate of sedimentation at Ireviken was the same as the average rate for the Silurian of Gotland, and if the Wenlock and Ludlow epochs each lasted 5 Ma, then the 4.5 m of strata between the first local extinctions and the change in clay mineralogy would correspond to 90 000 years. Rates of sedimentation on Gotland were not constant, however, and this number gives only an indication of the magnitude of time involved.

Elsewhere in the world the faunal change between the *P. am.* Z. and succeeding strata is equally dramatic (Table 9.1). All species with platform elements disappear, along with most of those with blade and ramiform elements and several with only coniform elements. Only about 25% of the species in the *P. am.* Z. are unaffected. Succeeding Silurian conodont faunas are characterized by about ten widespread lineages that persisted for most of the period. Hence there are greater differences between the faunas of the *P. am.* Z. and immediately younger Wenlock faunas than between the latter and, for example, those of the late Ludlow.

In the rest of the Wenlock, only *Kockelella patula*, briefly present in the *C. rigidus* Zone, has a typical platform element. Its ancestry is unknown, but the platform resembles those of some Llandovery taxa (e.g. *Aulacognathus*).

In the Ludlow, platform-equipped taxa reappear globally. The list includes the monotypic genera *Ancoradella* and *Polygnathoides*, species of Icriodontidae, *Kockelella variabilis*, and the poorly known '*Spathognathodus inclinatus*' *posthamatus*. These and other taxa form a 'mid-Ludlow' group, consisting of about half the number of conodont species known from the epoch (Jeppsson 1984). A particular feature is that the closest known relatives of many of these taxa are not from the late Wenlock, but in the *P. am.* Z. or adjacent strata. The *P. s.* Z. forms the later part and culmination of a Ludlow faunal episode in a similar way to the *P. am.* Z. in the late Llandovery and earliest Wenlock. Both zones coincide with episodes of atypical sedimentation in many areas, as does the *Kockelella patula* Zone (*K. p.* Z.) in a few places.

## 9.3 LITHOLOGIES OF THE *P. AMORPHOGNATHOIDES* ZONE

In the condensed limestone sequence of Silurian and Early Devonian age exposed at Cellon in Austria (Fig. 9.1) there are four thin units of shale with interbedded limestone within the uppermost Llandovery and the 21 m of Wenlock and Ludlow strata (Walliser 1964). The oldest starts in the Llandovery *P. celloni* Zone (2.37 m) and ends with two thin lime-

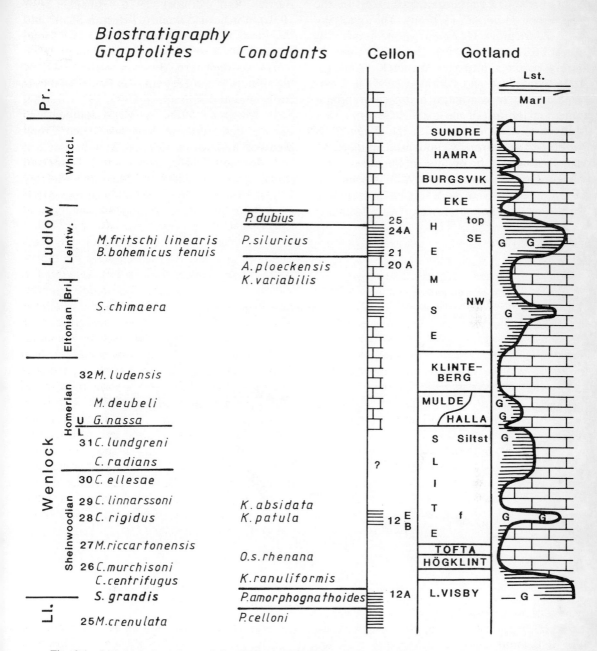

Fig. 9.1—Lithological sections of the Cellon and Gotland sequences, related to chronostratigraphical divisions, some graptolite zones and some conodont zones and faunas. The symbol 'G' in the Gotland column indicates levels where graptolites have been found. The marls on Gotland include both distal marls and 'lagoonal' marls (e.g. the Eke marl).

stone beds (0.27 m) forming the top of the succeeding *P. am.* Z. (1.13 m). The overlying *K. p.* Z. includes *C. rigidus* Zone (zone 28) graptolites (Jaeger 1975). Since the *P. am.* Z. barely reaches into the Wenlock (Aldridge 1975, Mabillard and Aldridge 1985) it is evident that there is a major hiatus (3 graptolite zones) at the top of the zone at Cellon.

In parts of North America the *P. am.* Z. is developed as a strongly condensed argillaceous unit, typically 1 m or less in thickness, with carbonates below and above. It has been recognized over a wide area (Fig. 9.2), including southern Oklahoma (Prices Falls Member;

Barrick and Klapper 1976, Amsden *et al.* 1980), northern Arkansas ('Button Shale' and the basal 0.3 m of overlying St Clair Limestone; Craig 1969, Barrick and Klapper 1976, p. 62), southeastern Missouri to the centre of the Illinois Basin (Seventy-six Shale Member; Satterfield and Thompson 1973, 1975, Barrick and Klapper 1976, p. 62), southeastern Indiana and adjacent Kentucky (Lee Creek Member and lower Osgood Member; Nicoll and Rexroad 1969), and central Tennessee (basal part of Maddox Member; Barrick 1983). However, more centrally on the continent, the *P. am.* Z. is developed largely as sev-

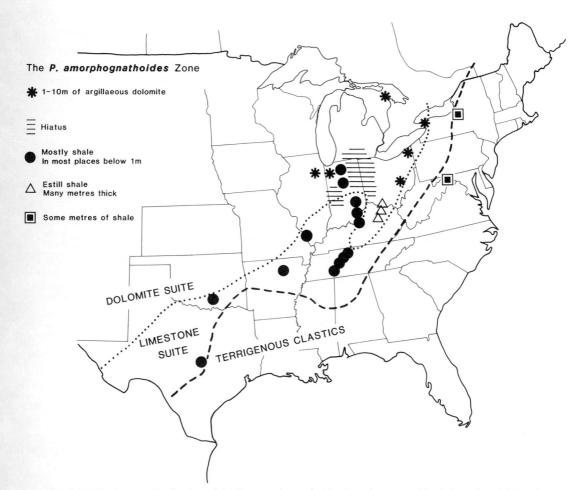

Fig. 9.2—The known distribution of the *P. amorphognathoides* Zone in eastern North America. A hiatus is marked only where it has been established by conodont studies. The distribution of the normal, background Silurian lithologies, as mapped by Berry and Boucot (1970), is indicated by broken lines.

eral (up to 10) metres of argillaceous carbonates, mainly dolomites. A thin shale horizon
occurs low in the zone in, for example, northeastern Illinois (Brandon Bridge Member;
Liebe and Rexroad 1977). Other areas include
northwestern Indiana (low Salamonie Dolomite; Rexroad and Droste 1982), eastern and
northeastern Ohio ('Packer Shell'; Kleffner
1985), the Niagara Gorge (Rockway Dolomite
Member; Rexroad and Rickard 1965), and the
Bruce Peninsula in Ontario (St Edmund
Member; Barnes *et al*. 1978).

More marginally on the North American
continent, argillaceous sediments dominate
below, in, and above the *P. am*. Z., which may
be more than 25 m thick (Rexroad and Nicoll
1972, fig. 1). Such lithologies occur in West
Virginia and Maryland (Helfrich 1980), and
probably in central New York State (Sauquoit
Shale; Rexroad and Rickard 1965, Berry and
Boucot 1970, p. 222: Willowvale Shale;
Rickard 1975). West of these, in eastern Kentucky and southern Ohio, the Estill Shale of
similar thickness occurs between carbonates
(Rexroad and Nicoll 1972).

Lateral tracing of the stratigraphical units
reveals that clastic sediments formed, instead
of the carbonates normal for the area,
throughout the Limestone Suite and in marginal areas of the Dolomite Suite (Fig. 9.2).
These changes have been ascribed to regional
conditions: a clastic tongue spreading from an
eastern source, thinning westwards and grading
through mixed lithologies into a carbonate unit
(Knight 1969, Rexroad and Nicoll 1972,
p. 59). The great thickness of the Estill Shale
agrees well with the interpretation that the
change from carbonates to argillaceous carbonate and calcareous shale was caused by dilu-
tion of the carbonate by an increased input of
clastics. However, elsewhere there are no indications that the thickness of the *P. am*. Z. was
increased by a greater rate of clastic sedimentation, nor that the number of conodont elements per kg was lowered because of the dilu-
tion. Instead the *P. am*. Z. is very thin (Nicoll
and Rexroad 1969, Craig 1969, Satterfield and

Thompson 1973, 1975, Barrick and Klapper
1976, Amsden *et al*. 1980, Barrick 1983), often
richer in conodont elements than the succeeding strata (e.g. Nicoll and Rexroad 1968, p. 5,
Liebe and Rexroad 1977, text-fig. 3), and contains glauconite and phosphate (Craig 1969,
Rexroad and Nicoll 1972, Satterfield and
Thompson 1975, Rexroad and Droste 1982,
Barrick 1983), minerals which are usually
interpreted as indicating a low rate of sedimentation. Further, the change is not regional,
but global. An alternative model is that the
regional changes in proportion between
carbonates and clastics originated in a strongly
lowered rate of calcium carbonate deposition.

Elsewhere (Fig. 9.3), the *P. am*. Z. has been
identified from the Canadian North-west Ter-
ritories (Chatterton and Perry 1983), Yukon
(M. J. Orchard, pers. comm.), southeastern
British Columbia (Norford 1976, Uyeno and
Barnes 1983, p. 10), southeastern Alaska
(Ovenshine and Webster 1969, 1970, Savage
1984), California (Miller 1976, 1978), central
Texas (Seddon 1970), Anticosti Island (Uyeno
and Barnes 1983), the Gaspé Peninsula (Nowlan 1981), and northwestern Greenland (Armstrong, in press). In Europe it is known from
Wales and the Welsh Borderland (Aldridge
1972, 1975, Aldridge and Mabillard 1981,
Mabillard and Aldridge 1982, 1983, 1985),
southern Norway (Aldridge 1974, Aldridge
and Mohamed 1982), Gotland (Jeppsson
1983), Estonia (Viira 1977, Klaamann 1984),
Lithuania (Saladžius 1975, Viira 1977),
Podolia (Drygant 1969), numerous localities in
Austria (for a summary see Schönlaub 1979,
1980), and Bosnia (Spasov and Filipović 1966,
Spasov 1966).

From Asia very few conodont records of this
age have been published, but the *P. am*. Z. has
been reported from northwestern Turkey
(Haas 1968), northwestern Malaysia (Igo and
Koike 1968), Tibet (Lin Bao-yu and Qiu
Hong-rong 1983, Lin Bao-yu 1983, Qiu
Hong-rong 1984), Japan (Kuwano 1976), and
Siberia (Mashkova 1977). It has also been
recorded from New South Wales, Australia

Fig. 9.3—Known global distribution of the *P. amorphognathoides* Zone, based only on localities where the zonal species has been found.

(Nicoll and Rexroad 1974, Bischoff 1975).

These localities are from several different Silurian continents, but there are too few from each to give as detailed a picture as that for eastern North America. However, most published lithological and/or thickness data for the *P. am.* Z. and adjacent strata fit the model developed on North American data, that the *P. am.* Z. coincides with a reduced rate of carbonate sedimentation in the marginal carbonate areas but unchanged conditions in more central and more offshore areas. For example, in northeast Turkey the *P. am.* Z. is developed as violet or rarely green shale rich in chamosite ooids, overlain by grey 'Flaserkalke' (Haas 1968, pp. 192–195). Considering different subsequent mineralization and weathering, these Turkish strata can be compared with those in northern Arkansas (Craig 1969), except that the beds are thicker. In Estonia the *P. am.* Z. consists of a thin argillaceous unit and is overlain by dolomitized limestone and dolomite (Klaamann 1984). The single sample described from Malaysia was from immediately above a black shale intercalated in a limestone sequence ranging from Lower or Middle Ordovician into Devonian (Igo and Koike 1968). In California the *P. am.* Z. makes up at least 42 m, possibly 55 m, of an unchanged dolomite sequence (Miller 1978), to date the maximum thickness reported.

In some areas, like parts of Greenland, carbonate sedimentation ceased with the *P. celloni* Zone (Armstrong in press, Hurst 1984). One possible explanation is that before the *P. am.* Z. Chron, the rate of deposition barely kept pace with the rate of subsidence, and that the uncompensated subsidence during the *P. am.* Z. Chron brought the sea bed below the zone of high carbonate sedimentation. Some lithological sequences, e.g. in the Welsh Borderland (Mabillard and Aldridge 1985), seem to fit a more complicated model.

## 9.4 LITHOLOGIES OF THE
## *P. SILURICUS* ZONE

The fourth shale episode in Cellon (Fig. 9.1) coincides exactly with the *P. s.* Z. (2.83 m; Walliser 1964). The *P. s.* Z. is similarly developed at several other localities in the Carnic Alps (Schönlaub 1970, 1980).

On Gotland, the *P. s.* Z. includes graptolitic shales (with *M. bohemicus*; Hede 1942) and marls to the west, marls in the marginal limestone area, and limestones in the most northeastern area (Jeppsson 1983).

In eastern North America the *P. s.* Z. is less well-known than the *P. am.* Z., but available data fit the *P. am.* Z. model. For example, the zone is represented by argillaceous limestone and shale in SE Missouri (Rexroad and Craig 1971), in the Limestone Suite area (Berry and Boucot 1970).

In Michigan, Ontario, and New York, the Pittsford Shale, a black shale at the base of the Salina Formation, typically about 0.3 m thick, shows conditions of slow sedimentation (Sandford 1972) and from the meagre palaeontological data may be referable to the *P. s.* Z., (Berry and Boucot 1970, pp. 179, 220; compare Rickard 1975).

In southeastern Cornwallis Island in the Canadian Arctic a 40 m thick tongue of dolomitic siltstone and limestone, with *M. bohemicus tenuis*, is intercalated in dolomites and limestones in the lower part of the *P. s.* Z., which is at least 207 m thick (Thorsteinsson 1981, p. 5, Thorsteinsson and Uyeno 1981, fig. 16).

In oceanic areas in the Palaeotethys, too, atypical sediments interrupt the graptolite shale sequence (Jaeger 1959, 1976, 1977, 1978). Thin intercalations of normal shale produce the only graptolites found in these units. In the deepest areas, e.g. Sachsen (Jaeger 1977) and the Carnic Alps (Jaeger 1976, 1977), about 5 m of grey-green shale

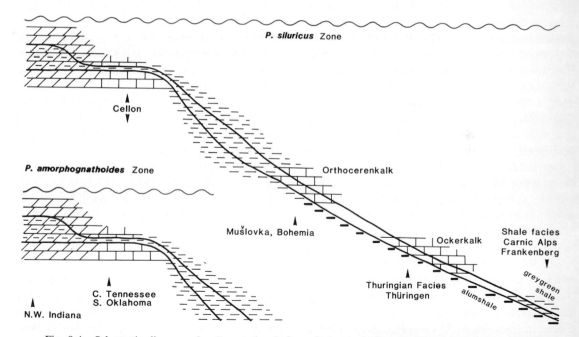

Fig. 9.4—Schematic diagram of sedimentation before, during and after the *P. amorphognathoides* and *P. siluricus* chrons. Some of the areas discussed in the text are marked. Thicknesses are not to a uniform scale, but relative thicknesses at each point indicate changes in the rate of sedimentation. Distance below sea level is not to scale.

interrupts the sequence of condensed graptolitic alum shale and lydite (Fig. 9.4). In the Thuringian Facies, 10–30 m of Ockerkalk interrupts a similar sequence in an area from Spain (near Barcelona) and Sardinia in the south, to Thüringen and Sachsen in the north and Czechoslovakia in the east (Jaeger 1976, p. 278). In both types of sequence it is the only episode with atypical sediments in the earliest Silurian–Early Devonian interval. In shallower (shoal) areas grey and black Orthocerenkalk is found, from Morocco and the Rheinische Schiefergebirge in the west to Asia in the east (Jaeger 1976). In Skåne, in southernmost Sweden, graptolite shale gives way to the Öved–Ramsåsa Group of various lithologies, such as calcareous mudstones and siltstones, limestone and sandstone, and with rich faunas but no graptolites (Jeppsson and Laufeld 1986). The youngest graptolite faunas below the atypical sediments vary between the base and the top of the *M. chimaera* Zone (Jaeger 1976). *M. fritschi linearis* is found in the lowest alum shale intercalations within the atypical sediments (Jaeger 1959, 1977). From Bohemia, conodont dates are available (Walliser 1964, Schönlaub 1980). In the Mušlovka Quarry, limestone beds are intercalated with shale except for a 14 m unit of biosparites ·(Schönlaub 1980). The final 1 m of shale below this unit contains *M. bohemicus tenuis* (see Schönlaub 1980). *M. fritschi linearis* has its highest occurrence about 3 m below the base of the biosparites (H. Jaeger, pers. comm., 1985). The *P. s.* Z. starts within the upper metre of shale and ranges 5 m into the limestone unit. The start of this widespread carbonate episode is close to the base of the *P. s.* Z. everywhere, while it continued well after the end of that zone.

## 9.5  OTHER ABIOTIC CHANGES

The presence of black shale in Britain together with evidence of sea level changes, fluctuations in the number of species of graptolites and

brachiopods, and other faunal data (Leggett *et al*. 1981), and $\delta^{34}S$ secular trends in Yukon (Goodfellow and Jonasson 1984), indicate that close to the base of the Wenlock, ocean conditions changed from stagnant, 'greenhouse', to more oxygenated. There is no black shale in Britain above the *M. riccartonensis* Zone except for a brief, weak episode in the early Ludfordian (Leggett *et al*. 1981, fig. 1), at the level of the *P. s.* Z. (but compare Leggett 1980, where the episode is placed in the early Gorstian).

The precise timing of the late Llandovery/early Wenlock event is difficult to ascertain, as the black shales are dated by graptolites, but no graptolites have as yet been described from the type locality for the base of the Wenlock (Bassett *et al*. 1975, p. 13). It is thus uncertain in which graptolite biozone the Llandovery/Wenlock boundary should be placed. There are a few published correlations between the graptolite and conodont zonations. Graptolites indicating a high level in the *M. crenulata* Zone (but below a *S. spiralis* level) have been found in the *P. celloni* Conodont Zone (Jaeger and Schönlaub 1970); and Norford (1976) and Uyeno and Barnes (1983) have reported conodonts of the *P. am.* Z. associated with a graptolite fauna indicative of the upper part of the *S. spiralis* Zone. In agreement with this, unpublished ranges of graptolites (R. Skoglund, pers. comm.) and conodonts in the När 1 core on Gotland indicate that both the *P. celloni* and *P. amorphognathoides* zones are found within the *S. spiralis* Zone (sensu Teller 1969) and that the top of the *P. am.* Z., and thereby also the base of the Wenlock as currently defined (Holland 1980) is well within that zone.

Leggett *et al*. (1981, fig. 1) show that there is a much lower frequency of black shales in the *M. crenulata* Zone (in their usage incorporating the *S. spiralis* Zone) than in the preceding Llandovery and the succeeding three Wenlock Zones. Thus, the *P. am.* Z. (and part or all of the *P. celloni* Zone) probably represents a brief spell of more oxygenated conditions.

Other published discussions of Silurian oceanic conditions have less precise time resolution and thus do not help in establishing the nature of the connection between variations in lithologies and in conodont faunas.

## 9.6 OTHER POSSIBLE SIMILAR EVENTS IN THE WENLOCK AND LUDLOW

There are different kinds of faunal and lithological evidence for the two events discussed above. In addition there are probably two, less well documented, and thus probably either weaker or briefer events in the time interval between them. Further, still more subtle, events may also be expected to have occurred.

The brief appearance of *Kockelella patula*, which has a true platform element, coincides with a shale unit (0.77 m) in Cellon. This interval was described as the *K. patula* Zone (*K. p. Z.*) by Walliser (1964).

In the eastern USA, *K. patula* is found high in the lower half of the *K. amsdeni* Zone (Barrick 1983, fig. 4). *K. absidata* replaces *K. walliseri* in the middle of that zone. On Gotland *K. walliseri* occurs in Slite Beds unit f and *K. absidata* in slightly younger beds (Jeppsson 1983). Slite f and contemporaneous strata on Gotland, which thus can be correlated with the *K. p. Z.*, fit the model developed for the *P. am. Z.* above. In marginal limestone areas the beds are developed as a thin, marly unit; closer to the shore-line limestone occurs, and more offshore the marls seem to have remained unaffected (Jeppsson 1983). *Monograptus priodon* became widespread on Gotland at that time, and occurs even in limestones; otherwise graptoloids are very rare on Gotland.

The third and thinnest shale episode in Cellon considered here is close to the base of the Ludlow. However, the conodont fauna studied from it is small (Walliser 1964), and at the present it cannot be related with any confidence to any lithological or faunal episode elsewhere.

## 9.7 CONCLUSIONS

Four similar episodes of unusual oceanic conditions are registered during the interval from the late Llandovery to the end of the Ludlow. They differ strongly in magnitude, the oldest and the youngest being the best developed, with the most distinct faunas, and with the best documented and probably also the most pronounced lithological deviations. The fact that conodont lineages that survived to reappear during the youngest episode studied did not appear during the preceding two episodes suggests that these were of a smaller magnitude. The much lower number of platform-equipped and other taxa during the youngest episode compared with the oldest does not in itself indicate that the youngest episode was weaker, since not all taxa are likely to have managed to escape extinction and to hold on in refuges until favourable oceanic conditions recommenced. However, most faunas representing the youngest episode are dominated by long-ranging taxa, with taxa of the 'mid-Ludlow-group' numerically subordinate; thus the youngest episode also deviated in some way from the oldest one documented here. In common to the oldest and youngest episodes are the facts that they both include more than one conodont zone, and that most taxa have locally varying, overlapping ranges. In common to all four is the reduction in carbonate sedimentation in areas where limestone was otherwise formed (Fig. 9.4). During the two strongest episodes the resulting starved sedimentation usually began contemporaneously with the final conodont zone; however, locally it started during the preceding zone. There is a very strong local aspect in the overlap of the taxa of the zones; this may indicate that the appearance of the zone fossil of the final zone was subject to local influence too. If so, then the lithological response may be a better clock.

Data from eastern North America for the oldest episode are sufficient to map the extent of the reduced carbonate sedimentation—the effect was felt from the edge of the carbonate shelf across the whole area of the Limestone Suite and also into the marginal Dolomite Suite. The more scattered data from other regions and from the later episodes are consistent with a similar course of events.

During the youngest episode, at least one ocean, stagnant during the preceding Silurian and again afterwards, became oxygenated and carbonate sedimentation began across it, except in the very deepest places. No carbonate blanket of similar extent developed during the preceding episodes in that ocean; however, data as yet not included in this model may indicate that local carbonate sedimentation was initiated within one or more of them. Oxygenated bottom conditions and carbonate sedimentation continued during a couple of zones after the carbonate-starved sedimentation had ceased in the epicratonic seas, except for some brief spells of normal stagnant sediments.

The oldest episode coincides with changes interpreted as indicative of oxygenated oceanic conditions, and the youngest episode coincides with at least partly similar conditions. The different conodont faunal response during the four episodes is not explained by a simple two-state model; however, the very rapid faunal change at the end of the two major episodes indicates that the gradual changes reached a threshold or a self-accelerating phase. Future conodont studies combined with geochemical studies with comparable high time resolution are likely to be important in achieving a refined model.

These episodes resemble some of those Devonian 'events' named by House (1985) in the disappearance of a number of taxa and in the presence of anomalous sediments. The differences are mainly interpretative: House did not identify any reappearances, and he discussed the possibility that the Devonian events may have had eustatic or euxinic immediate cause.

## 9.8  SOME PALAEOBIOLOGICAL CONSEQUENCES

Conodonts were very sensitive to oceanic changes. Most coniform-equipped taxa were either less sensitive than other taxa, or, more probably, reacted to other changes. A palaeobiological difference between some taxa with coniform elements and other taxa was discussed by Barnes and Fåhraeus (1975), who concluded that the former were pelagic while the others were nectobenthic.

Obviously, conodont lineages could exist without leaving a fossil record for at least during several million years; however, there is no reason to believe that the gaps in the record could not be much longer. The very large changes in population size and the likelihood that speciation would result if a species became entrenched in more than one isolated refuge would be expected to be important in the evolution of the group. The taxa discussed here were widespread only during the time of one or two zones, following a period of many zones when they were absent.

In spite of the fact that the time of absence probably is only about 10 million years or less, ancestral–descendant relationships remain to be established for most of the taxa discussed. Studies of such relationships in other groups are often based on ontogeny and on conservative structures. Fortunately much of the ontogeny of the conodont elements is preserved in the shape of internal structures, which have slowly gained increasing attention. Growth series, when available, give further information on ontogeny. Reconstruction of apparatuses permits use of the more slowly evolving non-platform elements in taxonomy, and that too is slowly gaining ground. Unfortunately, apparatus-reconstructions require much further work, especially in regard to rare taxa.

Most Silurian conodont zonal boundaries are based on the appearance and disapppearance of lineages, not on events within lineages, which generally are considered to be more reliable. The evidence presented here indicates that the former boundaries may also be highly reliable. Both kinds of correlation may theoretically be erroneous as surviving archaic populations and isolated 'premature' advanced populations can distort correlations based on events. One possible example of a minor event allowing a limited spread of a lineage is the local presence of *P. siluricus* in strata correlated with the *M. chimaera* Zone, as discussed above.

Identification of more recurrent sedimentary anomalies, conodont faunas, and lineages promises to be mutually beneficial for identifying the environmental requirements of different conodont taxa and for studies of oceanic changes, including identifying recurrent oceanic conditions. There is a limit to the resolution achievable using present conodont-based correlations, as recurrent conditions must be separated by at least a few million years to be safely established as not merely local fluctuations. Improved correlations are however possible (Shaw 1964, Sweet 1984).

## ACKNOWLEDGEMENTS

The quality of the manuscript was greatly improved by comments from Richard J. Aldridge, Stig M. Bergström, Ann-Sofi Jeppsson, Carl B. Rexroad, and an anonymous reviewer. Permission to cite unpublished data was given by Howard A. Armstrong, Hermann Jaeger, Michael J. Orchard, and Roland Skoglund. In discussions during ECOS IV, a number of people, including James D. Barrick, Brian D. E. Chatterton, Kirk Denkler, Michael A. Murphy, Rodney D. Norby, Michael J. Orchard, and Tom Uyeno, informed me about published and unpublished data that complement Fig. 9.2. Erna Hansson typed the manuscript and Christine Andréasson drafted the illustrations. My research is financed by the Swedish National Science Research Council directly and through Project Ecostratigraphy. To all my sincere thanks.

## REFERENCES

Aldridge, R. J. 1972. Llandovery conodonts from the Welsh Borderland. *Bulletin of the British Museum (Natural History) Geology* **22** 125–231, 9 pls.

Aldridge, R. J. 1974. An *amorphognathoides* Zone conodont fauna from the Silurian of the Ringerike area, south Norway. *Norsk Geologisk Tidsskrift* **54** 295–303.

Aldridge, R. J. 1975. The stratigraphic distribution of conodonts in the British Silurian. *Journal of the Geological Society* **131** 607–618, 3 pls.

Aldridge, R. J. 1976. Comparison of macrofossil communities and conodont distribution in the British Silurian. In: C. R. Barnes (ed.): *Conodont Paleoecology. Geological Association of Canada Special Paper* **15** 91–104.

Aldridge, R. J. and Jeppsson, L. 1984. Ecological specialists among Silurian conodonts. *Special Papers in Palaeontology* **32** 141–149.

Aldridge, R. J. and Mabillard, J. E. 1981. Local variations in the distribution of Silurian conodonts: an example from the *amorphognathoides* interval of the Welsh Basin. In: J. W. Neale and M. D. Brasier (eds.): *Microfossils from recent and fossil shelf seas*, Ellis Horwood, Chichester, Sussex 10–17.

Aldridge, R. J. and Mohamed, I. 1982. Conodont biostratigraphy of the early Silurian of the Oslo region. IUGS Subcommission on Silurian Stratigraphy. Field meeting, Oslo Region 1982. *Paleontological Contributions from the University of Oslo* **278** 109–120, 2 pls.

Alvarez, L., Alvarez, W., Asaro, F., and Michel, H. V. 1980. Extraterrestrial cause for the Cretaceous-Tertiary extinction. *Science* **208** 1095–1108.

Amsden, T. W., Toomey, D. F., and Barrick, J. E. 1980. Paleoenvironment of Fitzhugh Member of Clarita Formation (Silurian, Wenlockian) Southern Oklahoma. *Oklahoma Geological Survey Circular* **83** iii–iv + 1–54, 7 pls.

Armstrong, H. A. (in press). Conodonts from the early Silurian carbonate platform of North Greenland. *Grønlands geologiske Undersøgelse Bulletin*.

Barnes, C. R. and Fåhræus, L. E. 1975. Provinces, communities, and the proposed nektobenthic habit of Ordovician conodontophorids. *Lethaia* **8** 133–149.

Barnes, C. R., Telford, P. G., and Tarrant, G. A. 1978. Ordovician and Silurian conodont biostratigraphy, Manitoulin Island and Bruce Peninsula, Ontario. Geology of the Manitoulin Area. *Michigan Basin Geological Society Special Paper* **3** 63–71.

Barrick, J. E. 1983. Wenlockian (Silurian) conodont biostratigraphy, biofacies, and carbonate lithofacies, Wayne Formation, central Tennessee. *Journal of Paleontology* **57** 208–239.

Barrick, J. E. and Klapper, G. 1976. Multielement Silurian (Late Llandoverian-Wenlockian) conodonts of the Clarita Formation, Arbuckle Mountains, Oklahoma, and phylogeny of *Kockelella*. *Geologica et Palaeontologica* **10** 59–99, 4 pls.

Bassett, M. G., Cocks, L. R. M., Holland, C. H., Rickards, R. B., and Warren, P. T. 1975. The type Wenlock Series. *Institute of Geological Sciences Report* **75**/13 19 pp., 2 pls.

Berry, W. B. N. and Boucot, A. J. 1970. Correlation of the North American Silurian rocks. *Geological Society of America Special Paper* **102** 289 pp., 2 pls.

Bischoff, G. C. O. 1975. Conodonts. In: J. A. Talent., W. B. N. Berry, and A. J. Boucot (eds.): Correlation of the Silurian Rocks of Australia, New Zealand, and New Guinea. *Geological Society of America Special Paper* **150** 6–9.

Chatterton, B. D. E. and Perry, D. G. 1983. Silicified Silurian odontopleurid trilobites from the Mackenzie Mountains. *Paleontographica Canadiana* **1** 127 pp, 36 pls.

Craig, W. W. 1969. Lithic and conodont succession of Silurian strata, Batesville District, Arkansas. *Geological Society of America Bulletin* **80** 1621–1628.

Drygant, D. M. 1969. Conodonts from the Restevsky, Kitaigorodsky and Mukshinsky Horizons (Silurian of Podolia). *Paleontologiceskiij Sbornik* **6** 49–55, 1 pl.

Goodfellow, W. D. and Jonasson, J. R. 1984. Ocean stagnation and ventilation defined by $\delta^{34}S$ secular trends in pyrite and barite, Selwyn Basin, Yukon. *Geology* **12** 583–590.

Haas, W. 1968. Das Alt-Paläozoikum von Bithynien (Nordwest-Türkei). *Neues Jahrbuch für Geologie und Paläontologie Abhandlungen* **131** 178–242.

Hede, J. E. 1942. On the correlation of the Silurian of Gotland. *Lunds Geologiska Fältklubb 1982–1942*, 205–229. Also in *Meddelanden från Lunds Geologisk-Mineralogiska Institution* **101** 25 pp.

Hede, J. E. 1960. The Silurian of Gotland. In: G. Regnéll and J. E. Hede: The Lower Paleozoic of Scania. The Silurian of Gotland. *International Geological Congress XXI Session Norden 1960 Guidebook Sweden d. Stockholm*. Also in *Publications from the Institutes of Mineralogy, Palaeontology and Quaternary Geology, University of Lund, Sweden* **91** 44–89.

Helfrich, C. T. 1980. Late Llandovery-Early Wenlock conodonts from the upper part of the Rose Hill and the basal part of the Mifflintown Formations, Virginia, West Virginia, and Maryland. *Journal of Paleontology* **54** 557–569, 2 pls.

Holland, C. H. 1980. Silurian series and stages: decisions concerning chronostratigraphy. *Lethaia* **13** 238.

House, M. R. 1985. Correlation of mid-Palaeozoic ammonoid evolutionary events with global sedimentary perturbations. *Nature* **313** 17–22.

Hurst, J. M. 1984. Upper Ordovician and Silurian carbonate shelf stratigraphy, facies and evolution, eastern North Greenland. *Grönlands Geologiske Undersøgelse Bulletin* **148** 73 pp.

Igo, H. and Koike, T. 1968. Ordovician and Silurian conodonts from the Langkawi Islands, Malaya, Part II. Contributions to the geology and palaeontology of southeast Asia, XL. *Geology and Palaeontology of Southeast Asia* **IV** 1–21, 3 pls.

Jaeger, H. 1959. Graptolithen und Stratigraphie des jüngsten Thüringer Silurs. *Abhandlungen der deutschen Akademie der Wissenschaften zu Berlin Klasse für Chemie, Geologie und Biologie* **2** 227 pp., 14 pls.

Jaeger, H. 1975. Die Graptolithenführung im Silur/Devon des Cellon-Profils (Karnische Alpen). *Carinthia* **2** 111–126, 2 pls.

Jaeger, H. 1976. Das Silur und Unterdevon vom thüringischen Typ in Sardinien und seine regionalgeologische Bedeutung. *Nova Acta Leopoldina neue Folge* **45** Nr 224 263–299, 3 pls.

Jaeger, H. 1977. Das Silur/Lochkov-Profil im Frankenberger Zwischengebirge (Sachsen). *Freiberger Forschungshefte* C **326** 45–59, 1 pl.

Jaeger, H. 1978. Graptolithen aus dem Silur der Nördlichen Grauwackenzone (Ostalpen). *Mitteilungen der österreichischen geologischen Gesellschaft* **69** (1976), 89–107.

Jaeger, H. and Schönlaub, H. P. 1970. Ein Beitrag zum Verhältnis Conodonten-Parachronologie/ Graptolithen-Orthochronologie im älteren Silur. *Anzeiger der mathematisch-naturwissenschaftlichen Klasse der Österreichischen Akademie der Wissenschaften 1970* 85–90.

Jeppsson, L. 1983. Silurian conodont faunas from Gotland. *Fossils and Strata* **15** 121–144.

Jeppsson, L. 1984. Sudden appearances of Silurian conodont lineages—provincialism or special biofacies? In: D. L. Clark (ed.): Conodont biofacies and provincialism. *Geological Society of America Special Paper* **196** 103–112.

Jeppsson, L. and Laufeld, S. 1986. The Late Silurian Öved-Ramsåsa Group in Skåne, Southern Sweden. *Sveriges Geologiska Undersökning, Serie Ca* **57** 1–42.

Klaamann, E. 1984. Stop 5 : 4—Jädivere outcrop. In: D. Kaljo, E. Mustjõgi, and J. Zecker. (eds.): *International Geological Congress XXVII Session USSR Moscow 1984 Estonian Soviet Socialist Republic Excursions: 027 Hydrogeology of the Baltic 028 Geology and mineral deposits of Lower Palaeozoic of the Eastern Baltic area. Guidebook, Tallinn 1984*, 62.

Kleffner, M. A. 1985. Conodont biostratigraphy of the stray 'Clinton' and 'Packer Shell' (Silurian, Ohio subsurface) and its bearing on correlation. In: J. Gray, A. Maslowski, W. McCullough, and W. E. Shafer (eds.): *The New Clinton Collection—1985*, The Ohio Geological Society, Columbus, Ohio 219–230, 2 pls.

Knight, W. V. 1969. Historical and economic geology of Lower Silurian Clinton Sandstone of northeastern Ohio. *The American Association of Petroleum Geologists Bulletin* **53** 1421–1452.

Kuwano, Y. 1976. Finding of Silurian conodont assemblages from the Kurosegawa tectonic zone in Shikoku, Japan (In Japanese, with an English summary). *Memoirs of the National Science Museum, Tokyo* **9** 17–23, pl. 2.

Laufeld, S. 1974. Reference localities for palaeontology and geology in the Silurian of Gotland. *Sveriges Geologiska Undersökning* **C705** 172 pp.

Leggett, J. K. 1980. British Lower Palaeozoic black shales and their palaeo-oceanographic significance. *Journal of the Geological Society, London* **137** 139–156.

Leggett, J. K., McKerrow, W. S., Cocks, L. R. M., and Rickards, R. B. 1981. Periodicity in the Palaeozoic marine realm. *Journal of the Geological Society, London* **138** 167–176.

Liebe, R. M. and Rexroad, C. B. 1977. Conodonts from Alexandrian and Early Niagaran rocks in the Joliet, Illinois, area. *Journal of Paleontology* **51** 844–857, 2 pls.

Lin Bao-yu, 1983. New developments in conodont biostratigraphy of the Silurian of China. *Fossils and Strata* **15** 145–147.

Lin Bao-yu and Qiu Hong-rong 1983. The Silurian System in Xizang (Tibet) (In Chinese with an English summary). *Contributions to the geology of the Qinghai-Xizang (Tibet) Plateau* **8** 15–28.

Mabillard, J. E. and Aldridge, R. J. 1982. Arenace-ous foraminifera from the Llandovery/Wenlock boundary beds of the Wenlock Edge area, Shropshire. *Journal of micropalaeontology* **1** 129–136, 2 pls.

Mabillard, J. E. and Aldridge, R. J. 1983. Conodonts from the Coralliferous Group (Silurian) of Marloes Bay, South-West Dyfed, Wales. *Geologica et Palaeontologica* **17** 29–43, 4 pls.

Mabillard, J. E. and Aldridge, R. J. 1985. Microfossil distribution across the base of the Wenlock Series in the type area. *Palaeontology* **28** 89–100.

Mashkova, T. V. 1977. Novye konodonty zony amorphognathoides iz nizh-nego Silura Podolii. *Paleontologicheskij zhurnal* **4** 127–131.

McLaren, D. J. 1983. Bolides and biostratigraphy. *Geological Society of America Bulletin* **94** 313–324.

Miller, R. H. 1976. Revision of Upper Ordovician, Silurian, and Lower Devonian stratigraphy, southwestern Great Basin. *Geological Society of American Bulletin* **87** 961–968.

Miller, R. H. 1978. Early Silurian to Early Devonian conodont biostratigraphy and depositional environments of the Hidden Valley Dolomite, southeastern California. *Journal of Paleontology* **52** 323–344, 4 pls.

Nicoll, R. S. and Rexroad, C. B. 1969 [dated 1968] Stratigraphy and conodont paleontology of the Salamonie Dolomite and Lee Creek Member of the Brassfield Limestone (Silurian) in Southeastern Indiana and adjacent Kentucky. *Indiana Geological Survey Bulletin* **40** 73 pp., 7 pls.

Nicoll, R. S. and Rexroad, C. B. 1974. Llandovery (Silurian) conodonts from southern New South Wales. *Geological Society of America, Abstracts with Programs* **6** 534–535.

Norford, B. S. 1976. Faunas and correlation of the uppermost Lower Silurian Tegart Formation, southeastern British Columbia. *Geological Association of Canada, Mineralogical Association of Canada Program with Abstracts* **1** 39.

Nowlan, G. S. 1981. Late Ordovician–Early Silurian conodont biostratigraphy of the Gaspé Peninsula—a preliminary report. In: P. J. Lespérance (ed.): *Subcommission on Silurian Stratigraphy, Ordovician-Silurian Boundary working group. Field meeting, Anticosti-Gaspé, Québec 1981, Vol. II: Stratigraphy and Paleontology*, 257–291, 7 pls.

Odin, G. S., Spjeldnaes, N., Jeppsson, L., and Torshöj Nielsen, A. 1984. Possibilities of time scale calibration of the Silurian in diverse geological (P.T.) environments in Scandinavia. *Bulletin of liaison of I.G.C.P. Project 196* **3** 6–23.

Ovenshine, T. A. and Webster, G. D. 1969. Silurian conodonts from southeastern Alaska. *Geological Society of America, Abstracts with Programs for 1969, part 3*, 51.

Ovenshine, T. A. and Webster, G. D. 1970. Age and stratigraphy of the Heceta Limestone in northern Sea Otter Sound, southeastern Alaska. *United States Geological Survey Professional Paper* **700-C**, C170–174.

Qui Hong-rong 1984. Paleozoic and Triassic conodont faunas in Xizang (Tibet). (In Chinese, with an English abstract.) In: Li Guang-cen and J. L. Mercler (eds.): *Sino-French Cooperative investigation in Himalayas*, 85–112, 5 pls.

Rexroad, C. B. and Craig, W. W. 1971. Restudy of conodonts from the Bainbridge Formation (Silurian) at Lithium, Missouri. *Journal of Paleontology* **45** 684–703, pls 79–82.

Rexroad, C. B. and Droste, J. B. 1982. Stratigraphy and conodont paleontology of the Sexton Creek Limestone and the Salamonie Dolomite (Silurian) in northwestern Indiana. *Indiana Geological Survey Special Report* **25** 29 pp.

Rexroad, C. B. and Nicoll, R. S. 1972. Conodonts from the Estill Shale (Silurian, Kentucky and Ohio) and their bearing on multielement taxonomy. *Geologica et Palaeontologica*, **SB1** 57–74, 2 pls.

Rexroad, C. B. and Rickard, L. V. 1965. Zonal conodonts from the Silurian strata of the Niagara Gorge. *Journal of Paleontology* **39** 1217–1220.

Rickard, L. V. 1975. Correlation of the Silurian and Devonian Rocks in New York State. *New York State Museum and Science Service. Map and chart series number 24*, 16 pp.

Saladžius, V. 1975. Conodonts of the Llandoverian (Lower Silurian) deposits of Lithuania (in Russian, but with an English summary with this title). *Fauna i stratigrafiia Paleozoia i Mesozia Pribaltiki i Belorussii*, Vilnius. 219–226, 2 pls.

Sandford, J. T. 1972. Niagaran-Alexandrian (Silurian) stratigraphy and tectonics. In: R. T. Segall and R. A. Dunn (eds.): Niagaran stratigraphy: Hamilton, Ontario, *Michigan Basin Geological Society Annual Field Excursion—1972, 9 Sept 72—10 Sept 72*, 42–56.

Satterfield, I. R. and Thompson, T. L. 1973. Conodonts and stratigraphy of a shale unit Silurian) at the base of the Bainbridge Formation in southeastern Missouri. *Geological Society of America Abstracts with programs for 1973*, 348.

Satterfield, I. R. and Thompson, T. L. 1975. Seventy-Six Shale, a new member of the Bainbridge Formation (Silurian) in southeastern Missouri. *Missouri Geological Survey, Studies in stratigraphy: Missouri Geological Survey Depart-*

*ment of Natural Resources, Report Investigations* **57** 109–119.

Savage, N. M. 1984. Llandoverian-Wenlockian (Silurian) conodonts from southeast Alaska (Abs.). *27th International Geological Congress Abstracts* **1** 173.

Savage, N. M. 1985. Silurian (Llandovery-Wenlock) conodonts from the base of the Heceta Limestone, southeastern Alaska. *Canadian Journal of Earth Sciences* **22** 711–727.

Schönlaub, H. P. 1970. Vorläufige Mitteilung über die Neuaufnahme der silurischen Karbonatfazies der Zentralen Karnischen Alpen (Österreich). *Verhandlungen der Geologischen Bundesanstalt 1970*, 306–315, 2 pls.

Schönlaub, H. P. 1979. Das Paläozoikum in Österreich. *Abhandlungen der Geologischen Bundesanstalt* **33** 140 pp., 7 pls.

Schönlaub, H. P. (ed.) 1980. *Second European Conodont Symposium (ECOS II), Guidebook, Abstracts. Abhandlungen der Geologischen Bundesanstalt* **35** 213 pp., 25 pls.

Seddon, G. 1970. Pre-Chappel Conodonts of the Llano Region, Texas. *Bureau of economic geology, The University of Texas at Austin, Report of Investigations* **68** 130 pp., 19 pls.

Shaw, A. B. 1964. *Time in Stratigraphy*. McGraw-Hill, New York, N.Y. 365 pp.

Snäll, S. 1977. Silurian and Ordovician bentonites of Gotland (Sweden). *Acta Universitatis Stockholmiensis. Stockholm Contributions in Geology* **31** part 1. 80 pp., 9 pls.

Spasov, H. 1966. Significance of the conodont fauna for the stratigraphy of the Palaeozoic (in Russian). *Bulletin of the 'Strašimir Dimitrov' Institute of Geology* **15** 89–97, 1 pl.

Spasov, H. and Filipović, J. 1966. Konodontska fauna starijeg i mladeg Paleozoika ji i sz Bosne (The conodont fauna of the older and younger Palaeozoic in SE and NW Bosnia). *Geološki glasnik* **11** 33–54, 3 pls.

Sweet, W. C. 1984. Graphic correlation of upper Middle and Upper Ordovician rocks, North American Midcontinent Province, U.S.A. In: D. L. Bruton (ed.): *Aspects of the Ordovician System. Palaeontological Contributions from the University of Oslo* **295** 23–35.

Teller, L. 1969. The Silurian biostratigraphy of Poland based on graptolites. *Acta Geologica Polonica* **19** 393–501.

Thorsteinsson, R. 1981 (dated 1980). Stratigraphy and conodonts of Upper Silurian and Lower Devonian rocks in the environs of the Boothia uplift, Canadian Arctic Archipelago. Part 1 Contributions to stratigraphy. *Geological Survey of Canada Bulletin* **292** 1–38.

Thorsteinsson, R. and Uyeno, T. T. 1981 (dated 1980). Biostratigraphy. In: R. Thorsteinsson: Stratigraphy and conodonts of Upper Silurian and Lower Devonian rocks in the environs of the Boothia Uplift, Canadian Arctic Archipelago. Part I contributions to stratigraphy. *Geological Survey of Canada Bulletin* **292**, 21–31.

Uyeno, T. T. and Barnes, C. R. 1983. Conodonts of the Jupiter and Chicotte Formations (Lower Silurian), Anticosti Island, Québec. *Geological Survey of Canada Bulletin* **355** 49 pp., 9 pls.

Viira, V. 1977. Conodonts and their distribution in the Silurian of the East Baltic (Ohesaare, Kunkoiai, Ukmerge etc. borings) (in Russian). *Padii i Fauna Silura Pribaltiki*, Tallinn, 179–192.

Walliser, O. H. 1964. Conodonten des Silurs. *Abhandlungen des Hessischen Landesamtes für Bodenforschung* **41** 1–106, pls. 1–32.

# 10

# Cycles in conodont evolution from Devonian to mid-Carboniferous

W. Ziegler and H. R. Lane

## ABSTRACT

Pectiniform conodont elements occurring worldwide in rocks of Devonian to early Pennsylvanian age record at least seven evolutionary cycles. These cycles are: (1) late Silurian to Lochkovian; (2) Pragian to mid-Givetian; (3) mid-Givetian to Frasnian; (4) Famennian; (5) Tournaisian; (6) Viséan to early Namurian; and (7) post-mid-Carboniferous boundary. Each cycle, except for the last one, consists of successive low- and high-diversity episodes followed by an extinction event before returning to a low-diversity episode in the succeeding cycle. Generally, low-diversity episodes are characterized by the dominance of large-cavity pectiniform elements, whereas high-diversity episodes record evolutionary flowerings of pit-bearing forms. After loss of the pit-bearing pectiniform elements in the early Viséan, high-diversity episodes, such as those in earlier Devono-Mississippian time, never recurred, except perhaps in the Triassic. Instead, following major extinction events in post-early Viséan Palaeozoic and Triassic time, total taxonomic counts returned to pre-Viséan Devono-Mississippian low-diversity levels and remained there until the next major extinction event.

## 10.1  INTRODUCTION

Clark (1981, fig. 58; 1983, figs. 1–4) discussed variations in the rates of conodont generic evolution in mid-Cambrian to Triassic time. His graphs suggest that the origin and extinction of early Palaeozoic (Cambrian and Ordovician) conodont genera reached an impressive rate and that a tremendous diversity was attained during that time interval. According to Clark's work, many more conodont genera survived during the Cambro–Ordovician than became extinct. Toward the end of the Ordovician, however, his graphs show a very sharp decline below 'Index 1', that is the level at which surviving genera numbered the same as those becoming extinct. A period of relatively low-diversity was maintained through the Silurian and into the Devonian. A major recovery in conodont generic survival is dramatically illustrated for the Devonian and Mississippian, but thereafter generic extinctions mostly exceeded appearances and, as a consequence, the group tumbled from crisis to crisis until its final demise in the late Triassic.

By its nature, Clark's method of using genera for his documentation of originations and extinctions, and using the length of each geological series as the basic time increment within which generic counts were made can only produce very generalized conclusions about fluctuations in conodont diversity through time. Clark's graphs do, however, illustrate a very broad view of these variations in evolutionary rates that aid in a general understanding of the biological record by geological series, as well as providing a basis upon which more detailed studies below the generic level may proceed. For these reasons, we do not believe the diversity decrease at the end of the Ordovician to be as dramatic as Clark's graph suggests, as it is accentuated because of the long time periods utilized. Also, although an episode of relatively low-diversity was maintained during the Silurian, it was punctuated by two higher diversity episodes (i.e. *celloni-* + *amorphognathoides-* and *siluricus-*zones). Sweet's (1985, fig. 7) graph for Ordovician and Silurian conodont diversity was drawn using zones as the basic time interval and species as the taxonomic unit; it shows very similar diversity fluctuations to those presented herein for the Devonian and Mississippian (Figs 10.1, 10.2).

Our two graphs (Figs 10.1 and 10.2) demonstrate that quality of information from the plotting of species counts is dependent upon the fineness of the time increment used in graphing. The greater the time span for plotting specific counts, the more dramatic the fluctuations in the graph will appear. The Devonian graph (Fig. 10.1) is based on total counts within high- and low-diversity events that we define in this study. In contrast, Carboniferous counts (Fig. 10.2) are based on biostratigraphical zones and thus shorter time intervals. We believe that Fig. 10.2 illustrates a

more accurate and precise account of pectiniform element diversity fluctuations. In a study expanding further on this topic, we are developing a more systematic approach to

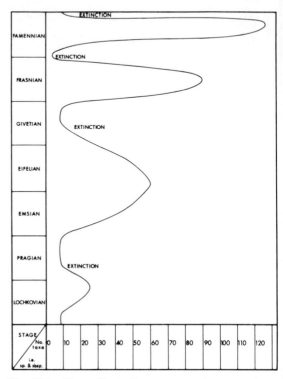

Fig. 10.1—Curve illustrating an approximate maximum number of pectiniform species within an entire episode rather than the exact number of species occurring per stratigraphical interval or zone. It employs the same technique used by Clark (1981, 1983) for creating his diversity graphs, except that this one is based on specific and subspecific rather than generic counts. The curve indicates several high- and low-diversity episodes in the Devonian, in contrast to that of Clark (1981, fig. 58, 1983, fig. 2), which illustrates one segment as a cumulative generic total for a series. The curve shows, in addition, that the maximum number of species within high-diversity episodes increases upwards in the Devonian.

Fig. 10.2—Chart illustrating the zone by zone variation in total species and subspecies counts for Mississippian to earliest Pennsylvanian (= Tournaisian–Namurian A) time. This chart differs significantly from Fig. 10.1 and that of Clark (1981, 1983) and is similar to that of Sweet (1985, Fig. 7) in that the basic time increment is the zone, rather than high- and low-diversity episodes (Fig. 10.1) or the series (Clark's graphs). It also differs from Clark's charts in that it is based on specific and subspecific rather than generic totals. Note that no distinct high-diversity episode occurs in the Viséan–Namurian interval after the early Viséan extinction event and before the mid-Carboniferous Boundary extinction event. Heavier black lines denote positions of extinction events. FU 8 to FU 10 are Faunal Units 8–10 of Lane (1974). The chart shows that there is a drop in diversity at the top of the Lower *crenulata-* Zone perhaps representing another cycle. [Note: the extinction at the bottom of the chart is within the lower part of the Middle *praesulcata-* Zone.]

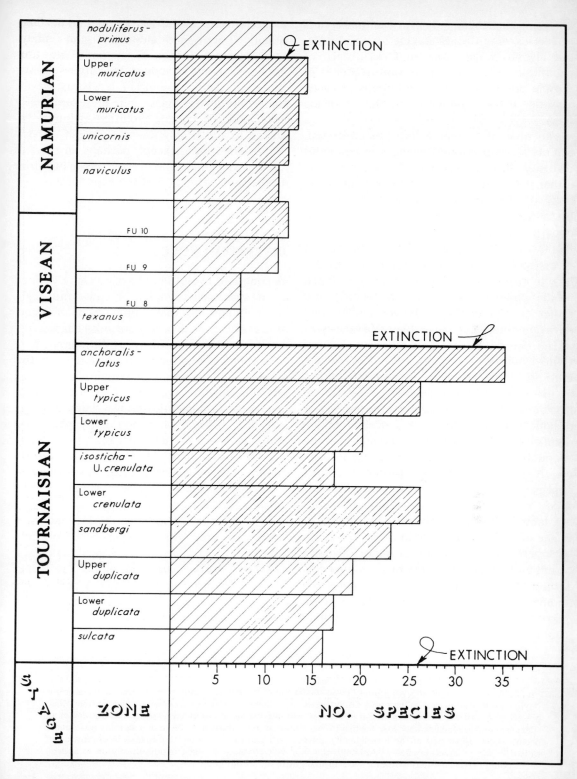

analysing each cycle, not only from total counts of species within zones making up an episode, but also from any consistent occurrence of pectiniform plans within the various phases of a cycle and episode (see definitions of cycle, episode, phase, and event in the following text).

Our qualitative approach in this paper represents a comparison of changes in conodont diversity through Devono–Mississippian time rather than an exhaustive analysis of the large body of data in our collections and in the literature. We describe the occurrence of conodont evolutionary cycles, consisting of high- and low-diversity episodes. The fluctuations in diversity (Fig. 10.1) seem not to be related only to eustatic sea level changes as developed in the paper by Johnson *et al*. (1985), but rather to a number of as yet unidentified environmental, tectonic and genetic factors.

The decline in conodont diversity in latest Ordovician time (see Clark 1981, 1983, Sweet 1985, p. 490–493) seems to correlate with a major climatic change leading to a series of glaciations that subsequently effected major oceanic changes. During this decline in the number of conodont species, the high-latitude conodont faunas known hitherto were completely eliminated and probably never appeared again in the conodont record (Sweet 1985, p. 493). Furthermore, contemporaneous low-latitude faunas underwent a major extinction event. This deep cut into the conodont biosphere was one of several major diversification and extinction events that continued to earliest Viséan time. Numerous new morphological groups were produced in these episodes of diversification (e.g., Polyg-

nathodontidae in the Early and Middle Devonian; Bactrognathodontidae and Idiognathodontidae in the Early and Late Carboniferous), and conodont diversity, at least at times, reached pre-Silurian levels. However, the evolutionary spark seen in several early and middle Palaeozoic evolutionary bursts discussed herein simply was not present in post-Tournaisian time, except, perhaps, in parts of the Pennsylvanian and in the Triassic just prior to the final extinction of the conodonts.

## 10.2   DEVONO-MISSISSIPPIAN DIVERSITY CYCLES

A conodont evolutionary cycle is here defined as a successive couplet of low- and high-diversity episodes that culminates in an extinction event. Low-diversity episodes characteristically have an innovative evolutionary phase near their closing. Innovative conodonts in this phase usually have a larger than normal basal opening and are the root stock for succeeding evolutionary flowering. High-diversity episodes display two evolutionary phases—a radiative phase in their early part, followed by a gradualistic phase. Thus, three evolutionary phases make up a normal cycle. An example of a complete cycle in the late Givetian to Frasnian is illustrated on Fig. 10.3.

Low-diversity episodes in the Devono–Mississippian are: (1) pre-*delta*-Zone Lochkovian; (2) pre-*dehiscens*-Zone Pragian; (3) mid-Givetian *varcus*-Zone; (4) pre-Upper *triangularis*-Zone Famennian; (5) post-Lower *praesulcata*-Zone Famennian and pre-*duplicata*-Zone Tournaisian; and (6) post-

Fig. 10.3—Example of an evolutionary cycle in the mid-Givetian to Frasnian time interval. Five major phylogenetic groupings are illustrated. Others exist, especially within the genera *Polygnathus* and *Icriodus*, but are not illustrated herein because they are less well understood and because of space limitation. Not all species names are included because the diagram is supposed to be illustrative. Note that an innovative phase occurs in the upper part of the low-diversity episode and that the high-diversity episode is comprised of radiative and gradualistic phases. Although this is a rather complete cycle, hypothetically an extinction event could occur anywhere within a cycle bringing it to an abrupt end.

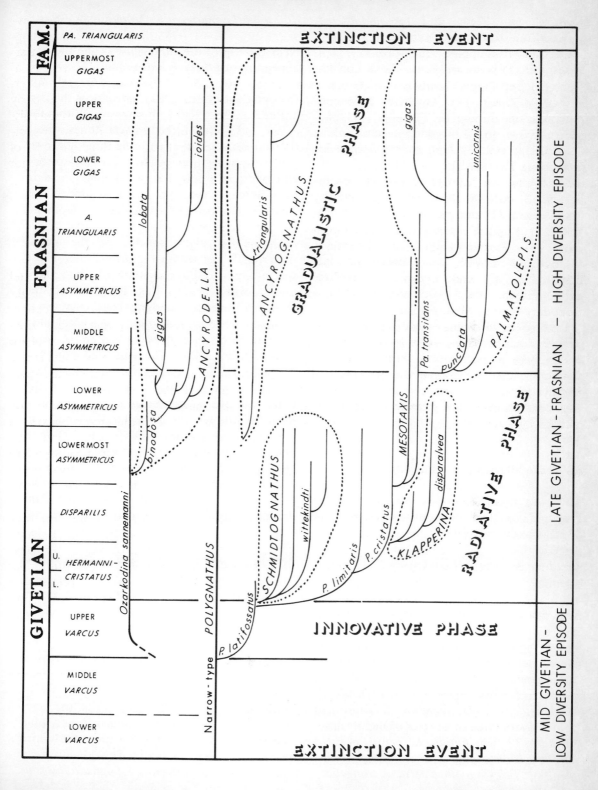

*anchoralis-latus*-Zone and pre-*noduliferus-primus*-Zone Viséan and Namurian.

Devono-Mississippian high-diversity episodes are: (1) post-*eurekaensis*-Zone Lochkovian; (2) late Pragian *dehiscens*- to Givetian *ensensis*- Zone; (3) Givetian *hermanni-cristatus*-Zone through the Frasnian; (4) Upper *triangularis*- to Lower *praesulcata*-Zone Famennian; and (5) *duplicata*- to *anchoralis-latus*-Zone Tournaisian.

Short-term or abrupt extinction events associated with the low- and high-diversity episodes are: (1) mid-*pesavis*-Zone late Lochkovian; (2) end of *ensensis*-Zone early Givetian; (3) Frasnian-Famennian Boundary; (4) end of Lower *praesulcata*-Zone late Famennian; (5) end *anchoralis-latus*-Zone early Viséan; and (6) mid-Carboniferous Boundary (end of Mississippian).

Each of the following episodes comprising a cycle is described using exclusively the known occurrences of pectiniform form-species and form-subspecies. Devonian events, episodes, and cycles, and their relation to the standard conodont zonation, are graphically illustrated on Figs 10.4 and 10.5.

### (a)    Late Silurian to Lochkovian cycle

This cycle began in post-*siluricus*-Zone Silurian time and continued to about the middle of the late Lochkovian *pesavis*-Zone.

### (i)    Late Silurian to Lochkovian low-diversity episode

This conodont low-diversity episode persisted into the early Lochkovian *woschmidti*- (*hesperius*) and *eurekaensis*-Zones. In these two zones, pectiniform taxa are represented by three genera including nine species and subspecies as follows: *Ozarkodina remscheidensis*, *O. paucidentata*, *O. repetitor*, *Icriodus woschmidti* (with three subspecies of slightly different ages, i.e. *woschmidti*, *hesperius*, *postwoschmidti*), *I. rectangularis*, *I. angustoides bidentatus*, and *Pelekysgnathus serratus elongatus*.

### (ii)    Late Lochkovian high-diversity episode

Near the top of the *eurekaensis*-Zone the first appearance of *Pedavis biexoramus*, *Ancyrodelloides omus*, and *Amydrotaxis sexidentata* (Murphy and Matti 1982), introduces the first episode of Devonian high diversity. Within the *delta*- and part of the *pesavis*-Zones, these three genera, together with the three holdover genera from the previous low-diversity episode and perhaps a few additional forms held in open nomenclature, produced 26 species in a spectacular short-term explosive development. Particularly important is the rapid gradualistic evolution of eight species of the genus *Ancyrodelloides* within the *delta*-Zone. These species are linked to one another by integradational forms. Within this development, the tendency to reduce the large basal cavity (e.g. in the oldest species, *A. omus*), to a basal pit in younger forms (e.g. *A. trigonicus* and *A. kutscheri*) is remarkable. The suggested evolution of these latter species (Murphy and Matti 1982) almost precisely matches that of the homeomorphic genus *Ancyrodella* in the Frasnian. Additional species of the two other genera are *Amydrotaxis johnsoni* and *Pedavis pesavis*. New appearances in *Ozarkodina* are *O. stygia* and *O. pandora*; those of *Icriodus* are *I. steinachensis*, *I. lotzei*, *I. alcolae*, *I. vinearum*, *I. fallax*, *I. simulator*, and *I. castilianus*. *Pandorinellina* is first represented by the species *P. optima*.

### (b)    Pragian to early Givetian cycle

This cycle began in the latest Lochkovian *pesavis*-Zone and continued to the base of the Givetian *varcus*-Zone.

### (i)    Early and middle Pragian low-diversity episode

In the early and middle Pragian (i.e. the highest *pesavis*-, *sulcata*-, and *kindlei*-Zones, nine species within six genera constitute the entire pectiniform element fauna. This low-diversity episode is characterized by *Eognathodus sul-*

| STAGE | ZONE | | EVENT | EPISODE | CYCLE |
|---|---|---|---|---|---|
| Frasnian | U'most | U'most | † EXTINCTION EVENT | | |
| Frasnian | U *gigas* | U | | HIGH | FRASNIAN |
| Frasnian | L | L | | | |
| Frasnian | *An. triangularis* | | | | |
| Frasnian | U | U | | — | — |
| Frasnian | M — *asymmetricus* — | M | | | |
| Frasnian | L | L | | | |
| Givetian | L'most | L'most | | DIVERSITY | |
| Givetian | *disparilis* | | | | MID- |
| Givetian | *hermanni-cristatus* | | | | GIVETIAN |
| Givetian | U | U | * INNOVATION | | |
| Givetian | M *varcus* | M | | LOW-DIVERSITY | |
| Givetian | L | L | | | |
| Eifelian | *ensensis* | | † EXTINCTION EVENT | | |
| Eifelian | *kockelianus* | | | | |
| Eifelian | *australis* | | | HIGH | EARLY |
| Eifelian | *costatus* | | | | GIVETIAN |
| Eifelian | *partitus* | | | | |
| Emsian | *patulus* | | | — | — |
| Emsian | *serotinus* | | | | |
| Emsian | *inversus -laticostatus* | | | DIVERSITY | |
| Emsian | *gronbergi* | | | | PRAGIAN |
| Pragian | *dehiscens* | | | | |
| Pragian | *kindlei* | | * INNOVATION | | |
| Pragian | *sulcatus* | | | LOW-DIVERSITY | |
| Lochkovian | *pesavis* | | † EXTINCTION EVENT | | |
| Lochkovian | *delta* | | | HIGH-DIVERSITY | LOCHKOVIAN |
| Lochkovian | *eurekaensis* | | * INNOVATION | | — |
| Lochkovian | *woschmidti /hesperius* | | | LOW-DIVERSITY | LATE SILURIAN |

Fig. 10.4—Pre-Famennian age events, episodes and cycles and their relation to the standard Devonian conodont zonation. †'s indicate extinction events and *'s indicate positions of introduction of innovative forms from which later radiative and gradualistic phases developed.

*catus*, *E. kindlei*, *Pandorinellina exigua philipi*, *Ozarkodina selfi*, *Icriodus steinachensis*, *I. claudiae*, *I. curvicauda*, *Pedavis mariannae*, and *Pelekysgnathus serratus*. Near the top of this interval *Polygnathus pireneae* occurs (Lane and Ormiston 1979, Murphy and Matti 1982, Chlupáč *et al.* 1985) forming the stock from which the important radiation of the genus *Polygnathus* developed.

### (ii)   Late Pragian to early Givetian high-diversity episode

The high-diversity episode associated with this cycle had its beginning in latest Pragian time at the base of the *dehiscens*-Zone and corresponds evolutionarily to the emplacement of the important Devono-Carboniferous genus *Polygnathus*. Successful morphological innovations in this case are the distinctive free blade, the well-outlined posterior platform with ridge and node ornamentation of the upper surface, and the establishment of the pit as a long-term stable morpholgical element of conodont gradualistic sequences. Earliest polygnathid species are, however, characterized by a large basal cavity beneath the whole of the platform. The gradualistic nature of *Polygnathus* speciation starting in the *dehiscens*-Zone continued through the Emsian, the Eifelian and into the early Givetian *ensensis*-Zone. The early evolution of the genus in the *dehiscens*- to *laticostatus*-Zones led to a reduction in size of the basal cavity (Klapper and Johnson 1975), until in the basal *serotinus*-Zone it first became a pit. At about that time, four to five polygnathid lineages came clearly into existence, each being variably characterized by special platform shape, ornamentation, blade-platform ratio, etc. This accelerated evolution of trends within *Polygnathus* is well documented in such studies as Klapper and Johnson (1975), Weddige (1977), Weddige and Ziegler (1979), Lane and Ormiston (1979), and others. The standard conodont zonation of this interval is based on species that came into existence during this period of gradualistic *Polygnathus* evolution. The boundary between the Lower and Middle Devonian Series is defined at the first occurrence of *Polygnathus costatus partitus*, a subspecies that appears as part of this gradualistic phase.

At the same time that the high-diversity episode was taking place in the more open marine environments of the *Polygnathus*-dominated faunas, a parallel development in the more restricted and endemic genus *Icriodus* was occurring (e.g. Weddige 1977, Weddige and Ziegler 1979). The majority of these late Early and early Middle Devonian icriodids developed from a small stock in the late Early Devonian (e.g. *I. corniger* Group) and then occupied the more geographically and environmentally restricted Middle Devonian shelves (Weddige and Ziegler 1976). However, only one or two icriodid species survived after the early Givetian extinction event.

As the result of this high-diversity episode, the *dehiscens* to *ensensis* interval is characterized as a whole by more than 60 species within such pectiniform genera as *Polygnathus*, *Icriodus*, *Pandorinellina*, *Sannemannia*, *Ozarkodina*, *Eognathodus*, and *Tortodus*.

### (c)   Mid-Givetian to Frasnian cycle

This cycle began at the base of the *varcus*-Zone and continued to the top of the Uppermost *gigas*-Zone.

### (i)   Mid-Givetian low-diversity episode

A period of lower diversity occurred in the mid-Givetian Lower and Middle *varcus* subzones. From the numerous polygnathid plans developed in the previous high-diversity episode, only two types remained (i.e. *Polygnathus linguiformis* and *P. varcus s.l.*, the latter including *P. timorensis* and *P. xylus xylus*) in the mid-Givetian (see Weddige 1977, text-fig. 4). *Icriodus brevis*, *I. difficilis*, *Ozarkodina brevis*, and *P. ansatus* are other associates of this time. *Ancyrolepis walliseri* and *Icriodus latericrescens s.s.* occur rather endemically.

Including *P. ovatinodosus*, which heralded a new type of ornamentation, about nine species existed in this low-diversity episode.

### (ii)    Late Givetian to Frasnian high-diversity episode

In the Upper *varcus*-Zone, *Polygnathus latifossatus* evolved from the *varcus* stock by developing a rather large basal cavity. *Ozarkodina sannemanni*, which has its first occurrence at this level, also has a broad basal cavity. Although iterative in regard to Early Devonian evolution, the large basal cavities in both these forms were once again an innovative morphological feature from which important subsequent rapidly-developing radiative evolutionary bursts arose. Following the emplacement of these two forms in the Upper *varcus*-Zone, a phase of radiation into new morphological types was initiated in the *hermanni-cristatus*-Zone. The time span up to the base of the Middle *asymmetricus*-Zone was a period of rapid radiation of many new morphological types, some resembling plans first developed in the late Lochkovian. The Middle–Upper Devonian series boundary has recently been defined at the base of the Lower *asymmetricus*-Zone within this radiative phase. Examples of these new morphological types follow.

*Schmidtognathus*: Five species evolved in the *hermanni-cristatus*-Zone (Ziegler 1966) from narrow polygnathids with large basal cavities.

Wide-platform *Polygnathus* (including *Mesotaxis*): Evolved from narrow polygnathids in the *hermanni-cristatus*-Zone. The first species are *P. limitaris* and *P. cristatus* in ascending order, then *M. asymmetricus* and *M. n. sp.* (former *P. ovalis*) (Ziegler 1966, Ziegler and Klapper 1982).

*Klapperina*: Evolved at the beginning of the *disparilis*-Zone from *P. cristatus* and includes the species *K. disparata*, *K. disparilis*, *K. dis-*

*paralvea*, *K. ovalis* (Ziegler *et al.* 1976, Ziegler and Klapper 1982).

*Palmatolepis*: Evolved at about the Lower *asymmetricus*-Zone from *P. asymmetricus* via *Palmatolepis transitans* and then into many species in the Frasnian (Ziegler 1962, Helms 1963, Helms and Ziegler 1981).

*Ancyrodella*: Evolved in the Lowermost *asymmetricus*-Zone from *Ozarkodina sannemanni* into *Ancyrodella binodosa* and in the Lower *asymmetricus*-Zone into *A. rotundiloba*. The latter is the marker for the beginning of the Upper Devonian and it continued the line on into many younger species (see Ziegler 1962, Klapper 1985).

*Ancyrognathus*: Evolved from *Polygnathus* into *A. ancyrognathoideus* at about the change from Lowermost *asymmetricus*- to Lower *asymmetricus*-Zone and then into typical lobed species (Ziegler 1962, Klapper and Lane 1985).

*Polygnathus*: Continuing from its flourishing in the Middle Devonian and near extinction in the mid-Givetian, *Polygnathus* recovered with many new species in the Frasnian (see Klapper and Lane 1985). A number of new *Polygnathus* species groups appeared and died out within the Frasnian including, among others, the *decorosus-*, *webbi-*, *brevis-*, *unicornis-*, and *angustidiscus*-groups. One of the several Frasnian conodont biofacies seemed to be populated dominantly by species of *Polygnathus* (see Klapper and Lane 1985). Although seemingly occurring in more restricted environments of deposition than the contemporaneous *Palmatolepis* biofacies, the relation of the *Polygnathus* biofacies to that characterized by *Icriodus* (Sandberg and Dreesen 1984), is not fully understood. Nevertheless, *Polygnathus* seems to be one of the most evolutionarily resilient pectiniform genera, having survived at least four low-diversity episodes and extinction events during Middle Devonian into Viséan time.

By about the base of the Middle *asymmetricus*-Zone, somewhat above the lower boundary of the Upper Devonian, most new morphological types were emplaced and from that level upwards, conodonts underwent a gradualistic development that produced numerous species throughout the Frasnian. The standard conodont zonation is based on species from these phylogenetic sequences (Ziegler 1962, 1971). Whereas most species of *Palmatolepis* are currently regarded as having inhabited more open marine areas, those of *Ancyrodella* and *Ancyrognathus* are

| STAGE | ZONE | | | EVENT | EPISODE | CYCLE |
|---|---|---|---|---|---|---|
| **FAMENNIAN** | U | *praesulcata* | U | * INNOVATION | LOW-DIVERSITY | TOURNAISIAN |
| | M | | M | † EXTINCTION EVENT | | |
| | L | | L | | | |
| | U | *expansa* | U | | | |
| | M | | M | | | |
| | L | | L | | | |
| | U | *postera* | U | | HIGH | **FAMENNIAN** |
| | L | | L | | | |
| | U | *trachytera* | U | | | |
| | L | | L | | — | |
| | U'most | | U'most | | | |
| | U | *marginifera* | U | | | |
| | L | | L | | | |
| | U | *rhomboidea* | U | | DIVERSITY | |
| | L | | L | | | |
| | U | *crepida* | U | | | |
| | M | | M | | | |
| | L | | L | | | |
| Frasnian | U | *Pa. triangularis* | U | * INNOVATION | LOW-DIVERSITY | |
| | M | | M | | | |
| | L | | L | EXTINCTION EVENT | | |

Fig. 10.5—Famennian age events, episodes and cycles and their relation to the standard Devonian conodont zonation. †'s indicate extinction events and *'s indicate positions of introduction of innovative forms from which later radiative and gradualistic phases developed.

found in somewhat more restricted environments.

### (d)  Famennian cycle

This cycle began at the base of the Lower *triangularis*-Zone and continued to within the Middle *praesulcata*-Zone of Ziegler and Sandberg (1984).

### (i)  Early Famennian low-diversity episode

A sharp extinction event occurred somewhat below the Frasnian-Famennian boundary (Fig. 10.6), that is, at the boundary of the Uppermost *gigas*- and Lower *triangularis*-zones (McLaren 1970, Ziegler 1984). At this position almost all previous species of palmatolepids, ancyrodellids, ancyrognathids, and polygnathids became extinct. A new species of

*Palmatolepis* (*Pa. triangularis*) is present in the Lower *triangularis*-Zone, probably having its ancestry within the Frasnian subgeneric manticolepid group. A short-lived low-diversity event in the Lower *triangularis*-Zone followed this major extinction event. Near the lower boundary of the Middle *triangularis*-Zone, the appearance of one new species of *Palmatolepis*, *Pa. delicatula*, and of one new species of *Icriodus*, *I. iowaensis*, heralded an innovative phase.

### (ii)  Mid-Famennian high-diversity episode

After the appearance of the innovative species, *Palmatolepis delicatula*, in the Middle *triangularis*-Zone, a rather abrupt radiative phase began at the lower boundary of the Upper *triangularis*-Zone. Many palmatolepid

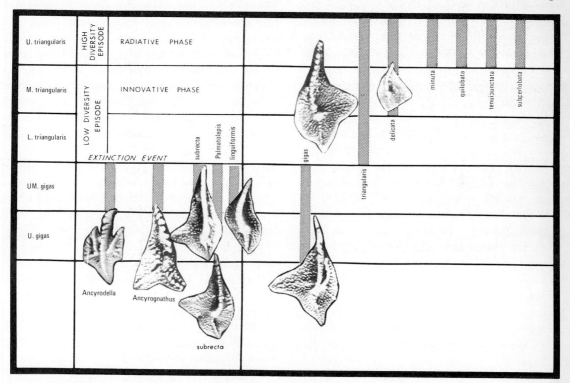

Fig. 10.6—Chart illustrating the extinction event in pectiniform conodonts near the Frasnian-Famennian Boundary. The extinction event is near the Uppermost *gigas* and *triangularis* zonal boundary. The low-diversity episode is within the Lower *triangularis*-Zone. The innovative phase corresponds approximately to the Middle *triangularis*-Zone and the radiative phase begins at the lower boundary of the Upper *triangularis*-Zone (Ziegler 1984). Two morphotypes of *Pa. subrecta* are illustrated.

(subgeneric group) species have their first occurrence at this level and gave rise to the well-known gradualistic phase on which the world standard conodont zonation of the Famennian is based (Ziegler 1962, 1971, Ziegler and Sandberg 1984). The four Famennian subgeneric evolutionary lineages of *Palmatolepis* species groups had their origin here and all produced morphological types that are unlike those in the Frasnian manticolepid species complex (Helms and Ziegler 1981).

The Famennian was also an episode of high diversity for *Polygnathus*. For example, the evolutionary profusion of the *nodocostatus*-group took place at this time (Helms 1963), as well as that of other *Polygnathus* species groups (i.e. *P. glaber*, *P. communis*, *P. semicostatus*, *P. brevilaminus*). Even the icriodontid development (*Icriodus*, *Antognathus*) accelerated somewhat so that Sandberg and Dreesen (1984) could establish a near-shore shallow-water zonation based on these species. *Clydagnathus*, *Patrognathus* and *Scaphignathus* are all genera that appear within the Famennian and whose distributions indicate near-shore biofacies.

The end of the Mid-Famennian high-diversity episode was heralded by a slow loss of species until only three palmatolepid species remained. They became extinct, along with species of *Polygnathus* and *Pseudopolygnathus*, just below the Devonian-Carboniferous Boundary in the Middle *praesulcata*-Zone (Ziegler and Sandberg 1984). Extinction also occurred in the majority of *Bispathodus* species that had just become established in the late Famennian. However, the early development of two genera (*Siphonodella* and *Protognathodus*) that were to become important in the Carboniferous took place at this time and continued upwards.

**(e)   Tournaisian cycle**

This cycle began within the Middle *praesulcata*-Zone and continued to the end of the *anchoralis-latus*-Zone in the early Viséan.

*(i)   Devonian-Mississippian boundary low-diversity episode*

Of all the profusion of Famennian species (more than 125 in total) only *Polygnathus communis*, three *Protognathodus* species, *Siphonodella praesulcata* and a *Bispathodus*, *Patrognathus*, and *Pseudopolygnathus* stock survived across this important stratigraphical boundary. The exact occurrence of the low diversity of conodonts is coincident with a mass extinction event during which many other fossil groups died out (clymenoids, some ammonoids, certain trilobites, and others, see Walliser (1983, 1984)).

*(ii)   Tournaisian high-diversity episode*

Towards the end of the Devonian-Mississippian boundary low-diversity episode, *Protognathodus kuehni* and *Siphonodella sulcata* appeared. The latter gave rise to an evolutionary flowering of *Siphonodella* species in the Tournaisian (Sandberg *et al.* 1978), which is the basis for the standard conodont zonation. At about the *duplicata*-Zone, most new morphological types came into existence by an accelerated radiative phase (Sandberg *et al.* 1978, fig. 1) that was followed by a period of gradualistic evolution during Tournaisian Tn1b to Tn3B (Kinderhookian) time. Other pectiniform genera that recovered after the Devono-Mississippian extinctions are *Polygnathus*, *Pseudopolygnathus*, and *Ozarkodina*. Whereas species of *Pseudopolygnathus* that became extinct in the latest Devonian were characterized generally by rather large basal cavities (e.g., *Ps. marburgensis*, *Ps. brevipennatus*), the *Pseudopolygnathus* species developing in the early Tournaisian (Kinderhookian) typically possessed smaller basal cavities (e.g., *Ps. marginatus*, *Ps. primus*).

In the course of the Tournaisian, there is a dramatic change in the composition of the pectiniform faunas. It began at about the base of the *isosticha*-Upper *crenulata*-Zone where the protognathodids (*Pr. praedelicatus*) gave rise to the important Mississippian genus *Gnathodus*. This is coincident with a marked decline and final extinction of the siphonodellids at the top of that zone. Conodont diversity was also significantly reduced at this time. Further study may show that this event subdivides the Tournaisian into two cycles, an early one ending at the top of the Lower *crenulata*-Zone and a late one ending at the top of the *anchoralis–latus*-Zone.

Diversity once again increased toward the top of the Lower *typicus*-Zone with the blossoming of gnathodids, pseudopolygnathids and the emplacement of the new family Bactrognathodontidae (*Bactrognathus*, *Dollymae*, *Doliognathus*, *Staurognathus*, *Eotaphrus*, and *Scaliognathus*). This high-diversity episode continued to the top of the succeeding *anchoralis–latus*-Zone.

The size reduction of the basal cavity in *Pseudopolygnathus* species observed in the early Tournaisian continued within the evolution of the *Pseudopolygnathus multistriatus* group (Lane *et al.* 1980, p. 123, fig. 3) and led to a *Polygnathus*-like shape and pit in *Ps. pinnatus*. The radiation of the numerous species of the Bactrognathodontidae in the late Tournaisian is a remarkable development, which is very helpful for stratigraphical correlation across many facies limits. This radiative event in the late Tournaisian is distributed worldwide and seems to be the last significant evolutionary burst of conodonts before their demise in the late Triassic.

This high-diversity episode, during which more than 40 pectiniform species existed, was brought to an end rather abruptly near the top of the *anchoralis–latus*-Zone by the extinction of most species. A very conservative stock of conodonts remained, including small numbers of gnathodids, ozarkodinids, eotaphrids and polygnathids, including *P. mehli*. In previous low-diversity or extinction intervals, forms like *P. mehli* (e.g. *P. pireneae*, *P. latifossatus*, *S. praesulcata*) carried on the genetic potential for evolutionary radiations of new forms exhibiting a pit. However, the *P. mehli* conser-

vative stock became extinct at about or slightly above the *anchoralis–latus/texanus* zonal boundary. Thus, by the time of the lower *texanus*-Zone, the ability to produce high-diversity blossomings associated with basal cavity to pit-forming evolutionary lineages was lost. We postulate that this loss of evolutionary potential resulted in a sudden narrowing of the genetic spectrum and was responsible for the low-diversity conodont faunas generally known from Viséan to early Triassic time.

**(f)   Viséan to early Namurian cycle**

This cycle began at the base of the *texanus*-Zone and continued to the mid-Carboniferous Bounday of Lane *et al.* (1986). By earlier Palaeozoic standards, the entire cycle is a low-diversity episode.

*(i)   Viséan to early Namurian low-diversity episode*

After the major extinction event at the end of the Tournaisian, a long period of low diversity was evident through the whole of the Viséan into the early Namurian. In terms of the North American succession, this includes the entire time span from the base of the *texanus*-Zone to the top of the late Chesterian Upper *muricatus*-Zone. This encompasses a number of low-diversity, long ranging faunas representing at least eight zones (= Faunal Units 7–14 of Lane and Ormiston 1982) including such species as *Gnathodus texanus*, *G. bilineatus*, *G. girtyi*, *G. pseudosemiglaber*, *Mestognathus beckmanni*, *M. bipluti*, *Taphrognathus varians*, *T. transatlanticus*, *Clydagnathus cavusformis*, *C. gilwernensis*, *Cloghergnathus globenskii*, *Cavusgnathus unicornis*, *C. altus*, *C. naviculus*, *Hindeodus scitulus*, *H. penescitulus*, *H. cristulus*, *H. spiculus*, *Ozarkodina campbelli*, *Adetognathus unicornis*, *Paragnathodus commutatus*, *P. homopunctatus*, *P. nodosus*, *Rhachistognathus prolixus*, and *R. muricatus*. Thus, an average of only three to four pectiniform species occurs in each Viséan to early

Namurian conodont zone. The top of this low-diversity episode corresponds to the top of the Mississippian in North America and to the mid-Carboniferous Boundary recently ratified by the Subcommission on Carboniferous Stratigraphy (Lane and Manger 1985).

## 10.3   POST-MID-CARBONIFEROUS BOUNDARY CYCLE

A very important extinction event rivaling those at the Ordovician–Silurian, Frasnian–Famennian, Devono–Mississippian, and Tournaisian–Viséan boundaries occurred at the Mississippian–Pennsylvanian boundary and involved not only conodonts, but also ammonoids, brachiopods, and foraminifers (see Ramsbottom *et al.* 1982). Until recently its importance for worldwide stratigraphical classification and correlation has been overlooked. At this level, species of the conodont genera *Cavusgnathus* and *Paragnathodus* disappeared, and the last species of the important Mississippian genus *Gnathodus* (e.g., *G. girtyi simplex* and *G. higginsi*) died out slightly above the main extinction event. At about the same time that these conodont genera were disappearing from the record, new species and genera characteristic of the Pennsylvanian appeared, marking the beginnings of the post-mid-Carboniferous cycle. Some examples are *Adetognathus lautus*, *A. spathus*, *Declinognathodus noduliferus*, *Rhachistognathus primus*, *R. websteri*, *R. minutus*, *Idiognathoides sinuatus*, *Neognathodus symmetricus*, and *Hindeodus minutus*. The base of this cycle is the level of the mid-Carboniferous Boundary of Lane *et al.* (1986) and is the same as the base of the Pennsylvanian in North America. The evolutionary succession represents a very modest recovery in diversity following the mid-Carboniferous Boundary extinction event, and in comparison with earlier recoveries already discussed, there is actually a slight decrease in total species count (Fig. 10.2). However, it represents one of a couple of

major changes in direction of conodont evolution in the Late Palaeozoic and Triassic. To term this new change in evolutionary direction a high-diversity episode would be incorrect. In fact, the zone-to-zone wide fluctuation in total species counts noted in pre-Viséan times simply does not seem to be developed in the Viséan to Triassic.

## 10.4   PATTERN OF DEVONO-CARBONIFEROUS CONODONT EVOLUTION

Our analysis of the fluctuations in conodont diversity within Devono–Carboniferous pectiniform species allows us to make some general statements. Conodont evolution in this interval seems to follow a cyclic pattern. An example of one of these cycles is illustrated on Fig. 10.3 and each can be subdividied into episodes, phases and events as follows:

(1) In the pre-Viséan Devono–Mississippian, a cycle begins with a short-term, low-diversity episode brought about by an extinction event. Generally, faunas in these low-diversity episodes are dominated by species having comparatively large basal openings, some of which gave rise to innovative new morphological features. Generally, species bearing such innovative features appear in the record near the end of a low-diversity episode.

(2) From these innovative forms, a short radiative phase developed, bringing about many new forms that adapted to and inhabited available ecological terrains. In the Devono–Mississippian, in particular, these radiative thrusts were coincident with the closing of the large basal opening of the innovative species to a pit. This radiation was the beginning of a high-diversity episode that was composed of two parts, a radiative phase followed by a gradualistic phase.

(3) A longer-term gradualistic phase, in which

a regular pattern of appearances and extinctions of species took place, followed the radiative phase. The first occurrences of new species within the gradualistic phases have been widely used for defining boundaries of conodont zones. In general, in the pre-Viséan Devonian and Mississippian, these gradualistic phases were dominated by pectiniforms that bear pits.

(4) An evolutionary cycle closed with the extinction of the last survivors of the high-diversity episode (Fig. 10.3), and a return to a low-diversity interval dominated by forms that bear large basal openings. These extinction events, although theoretically occurring at the end of a gradualistic phase, may actually occur anywhere within a cycle, bringing it to an abrupt end.

(5) High-diversity episodes, common in Late Cambrian to Tournaisian time, do not seem to have developed in post-Tournaisian time. Instead, recovery after major extinction events in post-Tournaisian time was very modest indeed, replacing extinct forms with approximately the same number of new taxa. Recovery in these cases was more a change in evolutionary direction as opposed to a burst of numerous new morphological types.

It is clear that each low-diversity episode followed an extinction event. Causes of these events are widely discussed, but still not known. Walliser (1984) has discussed the biological extinction near the Devonian–Mississippian Boundary, and concluded that global events had an impact on bioevolution. We believe that the results of our studies of conodont evolution support Walliser's conclusions. For example, Ziegler (1984) discussed the Frasnian–Famennian boundary and illustrated a black shale event at the time of extinction, which is similar to Walliser's conclusion in the case of the Devonian–Mississippian Boundary. Also, Sweet's (1985) end of Ordovician extinction event is linked to climatic changes that brought on a geologically sudden cooling of oceanic waters. However, we recognize that possibilities for biotic control in the form of abrupt radiations of predators, for example, need to be investigated and could have contributed to the evolutionary patterns of conodonts.

## ACKNOWLEDGEMENT

We thank Allen R. Ormiston, Amoco Research Center, for critically reading the manuscript and suggesting several modifications.

## REFERENCES

Clark, D. L. 1972. Early Permian crisis and its bearing on Permo-Triassic conodont taxonomy. *Geologica et Palaeontologica* **SB1** 147–158.

Clark, D. L. 1981. Biological considerations and extinctions. In: R. A. Robison (ed.): *Treatise on Invertebrate Paleontology, Part W, Supplement 2, Conodonta*. Geological Society of America and University of Kansas Press, Lawrence, Kansas, W83–W87.

Clark, D. L. 1983. Extinction of conodonts. *Journal of Paleontology* **57** 652–661.

Helms, J. S. 1963. Zur 'Phylogenese' und Taxionomie von *Palmatolepis* (Conodontida, Oberdevon). *Geologie* **12** 449–485, 4 pls.

Helms, J. S. and Ziegler, W. 1981. Fig. 62. In: R. A. Robison (ed.): *Treatise on Invertebrate Paleontology, Part W, Supplement 2, Conodonta*. Geological Society of America and University of Kansas Press, Lawrence, Kansas, W98–W99.

Jeppsson, L. 1985. Greenhouse conditions and conodont distribution. In: R. J. Aldridge, R. L. Austin and M. P. Smith (eds): *Fourth European Conodont Symposium (ECOS IV), Nottingham 1985, Abstracts*. University of Southampton, 14.

Klapper, G. 1985. Sequence in conodont genus *Ancyrodella* in Lower *asymmetricus* Zone (earliest Frasnian, Upper Devonian) of the Montagne Noire, France. *Palaeontographica, Abt. A* **188** 19–34, 10 pls.

Klapper, G. and Johnson, D. B. 1975. Sequence in conodont genus *Polygnathus* in Lower Devonian at Lone Mountain, Nevada. *Geologica et Palaeontologica* **9** 65–83, 3 pls.

Klapper, G. and Lane, H. R. 1985. Upper Devonian (Frasnian) conodonts of the *Polygnathus* biofacies, N.W.T. Canada. *Journal of Paleontology* 59 904–951.

Lane, H. R. 1974. Mississippian of Southeastern New Mexico and West Texas—A wedge-on-wedge relation. *Bulletin, American Association of Petroleum Geologists* 58 269–282.

Lane, H. R., Bouckaert, J., Brenckle, P., Einor, O. L., Havlena, V., Higgins, A. C., Jing-Zhi, Y., Manger, W., Nassichuk, W., Nemirovskaya, T., Owens, B., Ramsbottom, W. H. C., Reitlinger, E. A., and Weyant, M. 1986. Proposal for an International Mid-Carboniferous Boundary. *Compte rendu, 10th International Congress on Carboniferous Stratigraphy and Geology, Madrid* 4 323–339.

Lane, H. R. and Manger, W. L. 1985. The basis for a Mid-Carboniferous Boundary. *Episodes* 8 112–115.

Lane, H. R. and Ormiston, A. R. 1979. Siluro-Devonian biostratigraphy of the Salmon Trout River Area, East-Central Alaska. *Geologica et Palaeontologica* 13 39–96, 12 pls.

Lane, H. R. and Ormiston, A. R. 1982. Waulsortian facies Sacramento Mountains, New Mexico: Guide for an International Field Seminar, March 2–6, 1982. In: K. Bolton, H. R. Lane, and D. V. Lemone (eds.): *Symposium on the Paleoenvironmental Setting and Distribution of the Waulsortian Facies*. El Paso Geological Society and University of Texas at El Paso, 115–182, 3 pls.

Lane, H. R., Sandberg, C. A., and Ziegler, W. 1980. Taxonomy and phylogeny of some Lower Carboniferous conodonts and preliminary standard post-*Siphonodella* zonation. *Geologica et Palaeontologica* 14 117–164, 10 pls.

Mclaren, D. J. 1970. Presidential address: time, life and boundaries. *Journal of Paleontology* 44 801–815.

Murphy, M. A. and Matti, J. C. 1982. Lower Devonian conodonts (*hesperius-kindlei* Zones), Central Nevada. *University of California Publications in Geological Sciences* 123 1–83, 8 pls.

Ramsbottom, W. H. C., Saunders, W. B., and Owens, B. 1982. Biostratigraphic data for a Mid-Carboniferous Boundary. *Subcommission on Carboniferous Stratigraphy Publication* 8 1–156.

Sandberg, C. A. and Dreesen, R. 1984. Late Devonian icriodontid biofacies models and alternate shallow-water conodont zonation. In: D. L. Clark (ed.): *Conodont biofacies and provincialism*, *Geological Society of America Special Paper* 196 143–178, 4 pls.

Sandberg, C. A., Ziegler, W., Leuteritz, K., and Brill, S. M. 1978. Phylogeny, speciation, and zonation of *Siphonodella* (Conodonta, Upper Devonian and Lower Carboniferous). *Newsletters on Stratigraphy* 7 102–120.

Schönlaub, H. P. 1985. In: I. Chlupáč, *et al.*: The Lochkovian-Pragian boundary in the Lower Devonian of the Barrandian Area (Czechoslovakia). *Jahrbuch der Geologischen Bundesanstalt, Wien* 128 9–41, 4 pls.

Sweet, W. C. 1985. Conodonts: those fascinating little whatzits. *Journal of Paleontology* 59 485–494.

Walliser, O. H. 1983. Geologic processes and global events. *Terra Cognita* 4 17–20.

Walliser, O. H. 1984. Pleading for a natural D/C-Boundary. In: E. Paproth and M. Streel (eds.): *The Devonian–Carboniferous Boundary, Courier Forschungsinstitut Senckenberg* 67 241–246.

Weddige, K. 1977. Die Conodonten der Eifel-Stufe im Typusgebiet und in benachbarten Faziesgebieten. *Senckenbergiana Lethaea* 58 271–419, 6 pls.

Weddige, K. and Ziegler, W. 1976. The significance of *Icriodus*:*Polygnathus* ratios in limestones from the type Eifelian, Germany. In: C. R. Barnes (ed.): *Conodont Paleoecology, Geological Association of Canada Special Paper* 15 187–199.

Weddige, K. and Ziegler, W. 1979. Evolutionary patterns in Middle Devonian conodont genera *Polygnathus* and *Icriodus*. *Geologica et Palaeontologica* 13 157–164.

Ziegler, W. 1962. Taxionomie und Phylogenie Oberdevonischer Conodonten und ihre Stratigraphische Bedeutung. *Abhandlungen des Hessischen Landesamtes für Bodenforschung* 38, 1–166, 14 pls.

Ziegler, W. 1966 (date of imprint 1965). Eine Verfeinerung der Conodontengliederung an der Grenze Mittel-/Oberdevon. *Fortschritte in der Geologie von Rheinland und Westfalen* 9 647–676, 6 pls.

Ziegler, W. 1971. Conodont stratigraphy of the European Devonian. In: W. C. Sweet and S. M. Bergström (eds.): *Symposium on Conodont Biostratigraphy*. *Geological Society of America Memoir* 127 227–284.

Ziegler, W. 1984. Conodonts and the Frasnian/Famennian Crisis. *Geological Society of America, Abstracts with Programs* 16 no. 1, 73.

Ziegler, W. and Klapper, G. 1982. The *disparalis* Conodont Zone, the proposed level for the Middle-Upper Devonian Boundary. In: W. Ziegler and R. Werner (eds.):. *On Devonian stratigraphy and palaeontology of the Ardenno-Rhenish Mountains and related Devonian matters*. *Courier Forschungsinstitut Senckenberg* 55 463–492, 3 pls.

Ziegler, W. and Klapper, G. 1985. Stages of the Devonian System. *Episodes* **8** 104–109.

Ziegler, W., Klapper, G., and Johnson, J. G. 1976. Redefinition and subdivision of the *varcus*—Zone (Conodonts, Middle-?Upper Devonian) in Europe and North America. *Geologica et Palaeontologica* **10** 109–140, 4 pls.

Ziegler, W. and Sandberg, C. A. 1984. *Palmatolepis*-based revision of upper part of standard Late Devonian conodont zonation. In: D. L. Clark (ed.): *Conodont biofacies and provincialism. Geological Society of America Special Paper* **196** 179–194.

# 11

# Conodonts: the final fifty million years

D. L. Clark

## ABSTRACT

A species level analysis of the final 50 million years of the Conodonta defines three intervals, (1) a period of low diversity followed by, (2) a short interval of rapid diversification, and (3) another longer period of low diversification that culminates in extinction.

The Late Permian was a time of low diversification with no major extinction at the Permo-Triassic boundary. Early Triassic evolutionary rates were the highest of any time during the final 50 million years of the Conodonta. Originations and extinctions show some cyclicity during this final time of high diversity and rapid evolution. During the Middle Triassic, an interval twice as long as the Early Triassic, origination rates of species and genera were 50% lower than those of the Early Triassic. During the Late Triassic, only a single new genus evolved. The total of known Triassic species is approximately 100, distributed in at least 19 genera.

## 11.1 INTRODUCTION

Palaeontology currently enjoys a more lofty status in both scientific circles and the popular press than it has at any time since the 19th century 'dinosaur wars.' Propelled to glamor-

ous heights by the discussion of punctuated equilibria and what is called Darwinian gradualism, palaeontologists (or palaeobiologists as the new generation label themselves) have maintained considerable momentum by involvement in the discussion of mass extinction and extraterrestrial bodies and also through positive contributions to basic evolutionary theory. The Red Queen Hypothesis, coevolution, stasis, and periodicity of extinction have introduced concepts and terms that have become part of the necessary working vocabulary of all respectable palaeontologists, or at least, of palaeobiologists.

While conodont palaeontology has not figured prominently in the various public debates related to theoretical considerations of basic evolutionary principles, our speciality has shared at least a bit of the spotlight that has been captured by the profession as a whole because of the discovery of 'whole' conodonts with indications of soft parts, first in Scotland (Briggs *et al*. 1983) and now in Wisconsin (Mikulic *et al*. 1985). Also, there is the potential that proper interpretation of conodont element morphology may contribute to basic evolutionary theory.

The events surrounding the close of the conodont era are of concern here. The series of conodont originations, diversity changes, and extinctions have a distinct sequence that is similar to those described for other biological groups but with significant differences, as well.

For example, the general linearity of conodont survivorship curves can be considered as evidence that conodont evolution is understandable in the context of Van Valen's Law (Clark 1983). This 'Law' (Van Valen 1973) may be best explained by the Red Queen Hypothesis that suggests that organisms must evolve into better adaptive zones as quickly as possible just to survive because the host of predators, prey, and other ecologically related species that are part of the basic ecosystem are evolving new adaptive strategies at equal rates. Thus, species may become extinct at any time regardless of how long they have lived if evolutionary rates falter because, as the Red Queen predicts '. . . it takes all the running *you* can do, to keep in the same place.' (Lewis

Carroll, *Through the Looking Glass*). The results of all of this, according to the 'Hypothesis', are constant extinction rates for organisms. Probability of extinction does not increase or decrease during the range of a species at equilibrium diversity.

However, Wei and Kennett (1983), among others, interpret the linearity of survivorship curves used in Van Valen's Law only as evidence that longer-surviving species are not more resistant to extinction than their shorter-lived cousins, and that extinction rates of species are irregular and are affected by environmental (physical) perturbations through time yielding nonconstant rates of extinction, and hence, evolution (e.g. Stenseth and Maynard Smith 1984, Benton 1985, Fig. 11.1).

Raup (1984) reduced the entire process of

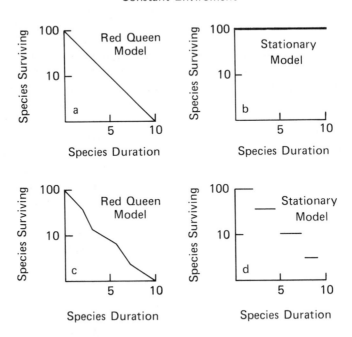

Fig. 11.1—Red Queen model of constant extinctions rates vs. 'Stationary' model of nonconstant extinction rates, under different environmental conditions. The Red Queen model has constant evolution rates with only slight variations at equilibrium diversity and in perturbed environments. The 'Stationary' model has zero rates of evolution at equilibrium but periodic bursts of evolution under perturbed environment. After Hoffman and Kitchell 1984, Stenseth and Maynard Smith 1984, and from Benton 1985.

speciation and extinction to a single equation,

$$S_t = S_0 e(p - q)^t$$

where p and q are speciation and extinction rates, $S_0$ is number of species present at some time $= 0$, and $S_t$ is number of species at time $= t$. Raup points out that exponential growth, decay, or stability depends upon $(p - q)$. Although extinction results from an increase in $q$, similar results come from a decrease in $p$ or anything that lowers the numerical value of $(p - q)$.

If conodont species numbers are plugged into Raup's evolutionary formula for the final 50 millon years, $(p - q)$ yields a negative number during the Middle Triassic, and this exponential decay results in Late Triassic (Norian) extinction.

An additional model for which conodonts may have some relevance has been proposed by House (1985), who reported a sequence of events in evolutionary patterns of Devonian ammonoids that may illustrate some basic evolutionary principles. His sequence includes (1) a decline in species diversity, followed by (2) cryptic appearance of novel groups, and (3) an increase in diversity, that leads to (4) morphological complexity and normal evolutionary behaviour of the group. The sequence is considered to be initiated by euxinic conditions, perhaps associated with global transgressive events.

The evolution of conodonts during their final 50 million years in part illustrates a similar cyclical pattern, but with some differences. For this report, origination, diversification, and extinction rates are merged into the encompassing 'evolutionary rates' terminology. Are any of the models important for conodonts? We need to begin with a look at the general pattern of conodont evolution.

## 11.2   CONODONT EVOLUTION

If the various 'protoconodonts' of the late Precambrian and Cambrian are considered to be ancestral to the later 'euconodonts' then the Conodonta survived for a little less than 400 million years. While this longevity is not remarkable for most phyla, the rapid evolution, biostratigraphical utility, rich diversification, widespread geographical distribution, and general abundance for the whole of the Palaeozoic and Triassic, distinguish conodonts as a significant group of fossils.

While a complete understanding of conodont evolution will never be derived from study of only the mineralized conodont elements, the morphological diversification of these elements has left patterns that can at least be described. Of particular interest is the pattern of diversification prior to extinction. Was the Norian demise of conodonts part of the major Triassic/Jurassic extinction event (Raup and Sepkoski 1984), precipitated by extraterrestrial or whatever ·cause, or was conodont extinction only a predictable part of their evolutionary history (Van Valen 1973)? Were extinction rates constant or variable? Conodont extinction is a microcosm of the whole picture of evolution and extinction that features so prominently in the literature today. Even though the data are limited, it is useful to determine if the evolutionary rates of conodonts are more similar to those predicted by the Red Queen Hypothesis and biotic interactions or by the Stationary Hypothesis and abiotic factors (Fig. 11.1). While a few details that describe the rise and fall of conodonts on the generic and family level are available (Clark 1980a, 1980b, 1983), this chapter addresses the evolutionary status of conodont species during the final 50 million years of conodont history.

## 11.3   THE DATA

For a group of organisms that has been studied for a relatively short time, it is remarkable that the taxonomy of Permian and Triassic conodonts is already so confusing. At the generic level there are fewer problems than at the species level (as is the case with most groups of

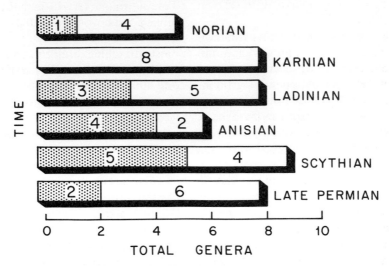

Fig. 11.2—Generic diversity during the final 50 million years of conodont history. Patterned area = number of new genera originating in that epoch; open area = number of holdover genera from previous epoch.

organisms), but it is already time for taxonomic monographs that will resolve differences in species interpretations. These monographs have not been published, and the data of this report are a subjective tabulation of the current literature tempered by an extensive experience with Permo—Triassic conodonts. I have accepted almost all reports of species and genera, and this totals approximately 100 species and 19 genera for the Triassic. While guaranteed not to be accurate, the numbers that are used are considered to be an accurate

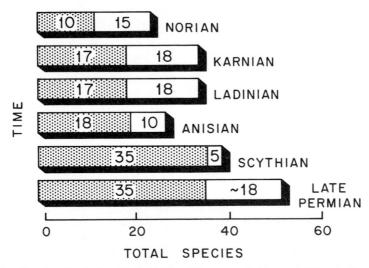

Fig. 11.3—Specific diversity during the final 50 million years. Patterned area = number of new species originating in that epoch; open area = number of holdover species from previous epoch.

approximation. Ages and dates used are based on the recent time scaling data of Harland *et al.* (1982) and Palmer (1983).

Figs 11.2 and 11.3 show generic and specific data for the final 50 million years of conodont evolution. There were at least 19 genera during the Triassic, probably six 'holdovers' from previous periods, and at least 13 new originations. Late Permian genera numbered at least eight, only two of which originated in the Late Permian. It is necessary to view originations (and extinctions) on a 'true time' basis (Fig. 11.4). The Late Permian represents approximately 13 million years. The Early Triassic (Scythian) is thought to represent only five million years. During the Scythian, there were at least nine genera, five of them new (the other four being holdovers from previous periods). There were at least 40 Scythian species, 35 of which originated during this epoch. The fact that there were five holdovers from the Late Permian has given rise to the idea that conodonts were relatively unaffected by the 'notorious P/T filter' (Sweet 1973). This idea is easily tested in terms of Late Permian–Scythian diversity.

## 11.4 PERMIAN TO EARLY TRIASSIC EVOLUTION

During the Late Permian (taken to represent some 13 million years) there were at least 50 species in eight genera of conodonts (Figs. 11.2, 11.3). During this 13 million year interval there probably were never more than approximately eight species living at one time. Most of the species living during the Late Permian originated during the Late Permian; two of the eight genera were new during that time. During the succeeding 15 million years (Scythian to Ladinian), there were approximately 70 species that originated and 12 new genera. Compared to the previous almost equivalent interval of time (Late Permian), the numbers of genera and species of the Early and Middle Triassic represent a 100% increase. This raises the question of whether the Late

Permian was an anomalously low interval of diversification or whether the Early and Middle Triassic were anomalously high intervals of diversification (at least as compared with the Permian). The former is apparently correct.

At least five Late Permian species are found in the earliest Triassic. The genus *Neospathodus* is reported in both the Late Permian and the Early Triassic, but I am not aware of an individual species of *Neospathodus* that occurs in both the youngest Permian and the Early Triassic (Permain species have been assigned to a different genus, *Merrillina*). The normal diversity of Late Permian conodont species was approximately five to six species, so the occurrence of this number on either side of the Permian/Triassic boundary confirms Sweet's (1973) conclusions (Figs 11.4, 11.5).

The size and diversity of Permian conodont faunas were constrained by an Early Permian extinction event (Clark 1972, Ritter 1985a,b). Late Early and later Permian faunal diversity is a reflection of an adjustment to this diversity reduction. Probably there were several minor expansion–reduction events during the remainder of the Permian, and the period cul-

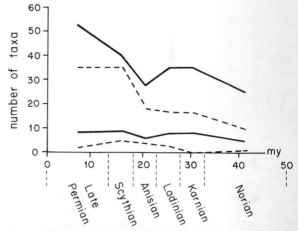

Fig. 11.4—Species and generic diversity, Late Permian to Norian, on true time plot. Upper pair of plots = specific diversity, solid line = total species living during that epoch, dashed line = total new species originating during that epoch. Lower pair = generic diversity, solid line = total genera living during that epoch, dashed line = total new genera originating during that epoch.

# PERMIAN                                                    TRIASSIC

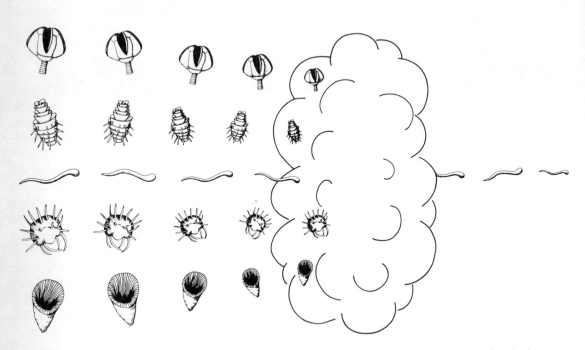

Fig. 11.5—Permo–Triassic boundary and extinction event. Conodonts survive this major time of extinction relatively unaffected while contemporaneous blastoids, fenestrate bryozoans, productid brachiopods, and rugose corals do not.

minated with a half dozen species of which at least five survived into the Triassic.

## 11.5  EARLY TRIASSIC EVOLUTION

Scythian evolution of conodonts began with as many genera and species as existed in the Late Permian (Fig. 11.4). Diversification was rapid in the Early Triassic, and during the five million years of this interval approximately 35 new species evolved, living with at least some of the five holdover species during the earliest part of this interval. Five generic originations occurred together with the four holdover genera. Of the four holdover genera, only two survived the Scythian and one survived into the Norian. The Scythian was the last interval of time that this level of diversification occurred. During succeeding and equal time intervals (Anisian,

Ladinian, Karnian) rates of diversification were only one half that of the Scythian. In fact, during the combined Anisian and Ladinian (=Middle Triassic) there were seven new genera and approximately 35 species originations, almost the same number as during the Scythian, but the Middle Triassic was evidently at least twice as long (Figs. 11.2, 11.3). It took 10 million years of Middle Triassic evolution to do what was achieved in five million years of Early Triassic time.

Evolution during the Scythian was cyclic and there was specific diversity reduction at each of the Scythian age boundaries (Fig. 11.6). The diversity reduction that followed the Griesbachian resulted in a smaller Dienerian holdover stock of species, and new species originations returned the Dienerian conodont diversity to Griesbachian levels. Smithian conodonts exhibit the greatest diversification of the

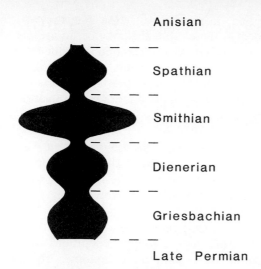

Anisian

Spathian

Smithian

Dienerian

Griesbachian

Late Permian

Fig. 11.6—Cyclic species diversification and reduction during the Scythian (Early Triassic). Width of figure indicates relative diversity. Notice diversification of species at each age boundary and high Smithian diversification. If these Scythian ages are of approximately equal duration, then this cycle of diversification to reduction covers approximately 1.2 million years. Data based primarily on Svalbard and Utah Scythian sections but this approximates ocean-wide conditions.

Scythian, including many novel species that are very short-ranging (Carr *et al*. 1984). The Smithian–Spathian reduction left the Spathian conodont diversity levels at the same threshold as those of other age boundaries. Spathian species numbers returned to early Scythian levels and were similarly reduced to begin the Anisian (Middle Triassic).

If the Scythian represents only five million years, then this cyclic expansion–reduction of species numbers may have occurred at approximate 1.25 million year intervals, but this is based on an unproved assumption that the Scythian substages are of equal length. Nonetheless, this pattern appears unique for Triassic conodonts. The Smithian diversification that included several short-ranging novel species and genera occurred approximately 26 million years after the Early Permian 'post-crisis' diversification (Ritter 1985a). It does not appear possible to relate this 26 million year cycle with that reported by Raup and

Sepkoski (1984) and used as the basis for the various Earth impact models, but the unique species produced at the beginning and end of this conodont cycle are of considerable significance in conodont biostratigraphy (Ritter 1985a).

The problem with all of these plots is the crudeness of the time scale. Exactly during what part of the Scythian ages did the species become extinct? If all of the species became extinct at the same time, a different interpretation could be expected than if the extinction rates were random during an age. Similarly, if extinctions are concentrated in a part of the age (something to be expected if there were a sudden environmentally related cause), rates of extinction per million years lose their meaning (Raup 1984). Also, the precise time of originations is obviously affected by this same consideration, and if all species evolved within a small time interval, the rate of evolution per million years is meaningless.

## 11.6　MIDDLE TO LATE TRIASSIC EVOLUTION

Following the Early Triassic diversity increase, there was an abrupt decrease in the early Middle Triassic. During the Anisian approximately two-thirds as many species were present as had existed during the Scythian, and approximately one-third of these were holdovers from the Scythian. Species originations were reduced by 50% even though generic originations remained almost constant. Ladinian species increased, but only partly because of new originations; more than one-half of all Ladinian species (about 18) were holdovers from the Anisian. As earlier suggested, total Middle Triassic evolutionary rates were approximately one-half of those of the Early Triassic. The combined Anisian–Ladinian evolutionary pattern is one of leveling out, a sort of *status quo* for conodonts, down from the levels of Scythian activity, and poised for the final curtain. No significant Anisian–Ladinian boundary diversity crises such as those of the Scythian ages (Fig. 11.6) are apparent. It appears that there may never have been more than about eight species living at the same time.

The most profound change in evolution rates occurred during the Late Triassic. While approximately the same number of species originated during the Karnian as in the preceding and equivalent intervals of time (5 million years), there evidently were no new generic originations. The Norian (including Rhaetian) continued this trend; only ten new species and one new genus originated during a 17 million year interval (as long as the preceding Anisian, Ladinian, and Karnian combined). Put another way, during the Late Triassic (Karnian and Norian = 22 million years), there were 27 species originations and one genus origination, but in the previous 15 million years (Scythian, Anisian, and Ladinian) there were 63 species and 12 generic originations. Survivorship curves for generic and family data illustrate the linear nature of this trend (Clark 1983). Clearly, the most significant change in evolutionary patterns from the Early Triassic was absence of new generic originations. During the final 50 million years of their evolution, conodonts had mean speciation rates of 2.8 species/my and generic originations of 0.3 genera/my. The figure for only the Triassic part of the final 50 million years is approximately the same. Species and genera originations per my are greatly reduced during the Late Triassic.

## 11.7  EXTINCTION

Details of the extinction of conodonts derived from studies of Triassic/Jurassic marine sections in Austria and Nevada have been published elsewhere (Clark 1983). Conodonts did not survive the latest Norian (Rhaetian of earlier terminology), although there were at least four species living during that time. There is some suggestion that conodonts may have survived slightly longer (into the very latest

Norian) in the Tethyan area than they did in North America, but this may be an artifact of poor correlation of series and systemic boundaries between continents. Abundance was greatly reduced during the latest Norian, and the final Tethyan conodont fauna consisted of only four species in very small populations (Clark 1983).

The extinction of the Conodonta in Austria and Nevada occurred at the same time as an environmental change. The carbonate rocks in the two areas indicate that low-energy, perhaps basinal, conditions (with a pelagic fauna including conodonts) changed and was replaced with shallower, higher-energy conditions with a benthic fauna and no conodonts. This was not an environmental change of great magnitude. Similar changes during the Palaeozoic and Triassic were easily handled by conodonts. Perhaps the four small populations of conodonts became extinct at about the same time for very different reasons, none of which is understood (Clark 1983).

## 11.8  SUMMARY AND CONCLUSIONS

Assuming the accuracy of the Triassic ages and of the specific and generic data obtained by a subjective review of Triassic literature, the following conclusions are made:

(1) Late Permian conodont evolutionary rates were relatively low, approximately one new generic origination per six million years and approximately 2.7 species per my. This interval of low diversity resulted from the Early Permian taxonomic pruning.

(2) Scythian evolution was more rapid, one new generic and 7.6 species originations per my. This was the last major surge of conodont evolution.

(3) Anisian evolution was one-half that of the previous epoch; 0.8 genera per my and 3.6 species per my.

(4) The Ladinian pattern was approximately

the same as that of the Anisian; 0.6 genera and 3.4 species per my.

(5) Karnian evolution shows a remarkable change in generic originations, none in five million years. Evolution could not be maintained by slowing down. At the species level, the same rate as that of the Ladinian held, 3.4 species per my.

(6) The Norian Epoch was a time of extinction. During this 17 million year interval (almost as long as the combined pre-Norian Triassic), there was a single new genus origination, 0.05 per my, and 0.8 species per my. The Norian is three times as long as any other Triassic epoch, but during this time only $\frac{1}{4}$ as many species and $\frac{1}{5}$ as many generic originations occurred as in the Scythian. There were approximately five species of conodonts living in the Norian seas, and all became extinct before the close of the Triassic.

(7) Perhaps the most significant evolutionary threshold of the final 50 million years occurred at the Middle/Late Triassic boundary. After this time, only one new genus and 27 new species originated (in 22 million years), while 12 new genera and 63 new species had originated in the previous 15 million years

A comparison of Triassic conodont species-level evolutionary patterns with patterns shown by other groups is not really possible, for the simple reason that most students are reluctant to be as 'objectively subjective' as is necessary to produce the data. Conodonts clearly show periods of rapid diversification terminated by periods of extinction, just as has been shown in other organisms (e.g. House 1985); but in the Triassic, this cycle apparently was completed in less than two million years (Fig. 11.6) and on a much reduced level of activity. The appearance of novel genera and species during the middle Scythian was not repeated until the Late Triassic, and then on a different level. The morphological variety of the Scythian is of the same magnitude as that of

the late Early Permian, and these two events are separated by an interval of approximately 26 million years.

Overall, conodont generic and family level evolutionary patterns are consistent with the model exhibited by the Red Queen Hypothesis (Clark 1983), although the data are so few that a comparison is quite subjective. On the species level, there is less consistency in evolutionary rates, however. The Late Permian evolutionary rates are greatly altered during the Scythian. The following Anisian to Karnian interval has consistent rates, but these are much lower than those of the Scythian. Thus, the perturbations of the subjective specific evolutionary rates may be interpreted to support the nonconstant model of evolution (Stenseth and Maynard Smith 1984, Wei and Kennett 1983), while the overall picture derived from a generic level analysis (which may be more accurate), could be interpreted to support the Red Queen Hypothesis (Fig. 11.1).

## ACKNOWLEDGMENTS

I have reviewed almost 500 Triassic and Permian publications for this compilation. This is essentially all of the Late Permian–Triassic literature published, almost all of it in the past 30 years. I acknowledge all of the authors who have described these species. This includes graduate students who produced 15 theses that have been of great assistance to my Permo–Triassic research during the past 20 years. These students include L. Cameron Mosher, Fred H. Behnken, Laurel C. Babcock, Mark A. Solien, John A. Larson, William A. Morgan, Rachel K. Paull, Tim R. Carr, Eric W. Hatleberg, Reed H. Meek, Ronald R. Charpentier, Stephen P. Carey, and Scott M. Ritter. The National Science Foundation has supported most of our work, and this includes current funding from EAR–8205675.

## REFERENCES

Benton, M. J. 1985. The Red Queen put to test. *Nature* **313** 734–735.

Briggs, D. E. G., Clarkson, E. N. K., and Aldridge, R. J. 1983. The conodont animal. *Lethaia* **16** 1–14.

Carr, T. R., Paull, R. K., and Clark, D. L. 1984. Conodont paleoecology and biofacies analysis of the Lower Triassic Thaynes Formation in the Cordilleran Miogeocline. In: D. L. Clark (ed.): *Conodont biofacies and provincialism, Geological Society of America Special Paper* **196** 283–293.

Clark, D. L. 1972. Early Permian crisis and its bearing on Permo-Triassic conodont taxonomy. *Geologica et Palaeontologica* **SB1** 147–158.

Clark, D. L. 1980a. Rise and fall of Triassic conodonts (abstract). *American Association of Petroleum Geologists Bulletin* **64** 691.

Clark, D. L. 1980b. Extinction of Triassic conodonts. In: H. P. Schönlaub (ed.): *Second European Conodont Symposium (ECOS II), Guidebook, Abstracts, Abhandlungen der Geologische Bundesanstalt, Wien* **35** 193–195.

Clark, D. L. 1983. Extinction of conodonts. *Journal of Paleontology* **57** 652–661.

Harland, W. B., Cox, A. V., Llewellyn, P. G., Pickton, C. A. G., Smith, A. G., and Walters, R. 1982. *A geologic time scale*. Cambridge Earth Science Series, Cambridge University Press, 131 pp.

Hoffman, A. and Kitchell, J. A. 1984. Evolution in a pelagic planktic system: a paleobiologic test of models of multispecies evolution. *Paleobiology* **10** 9–33.

House, M. R. 1985. Correlation of mid-Palaeozoic ammonoid evolutionary events with global sedimentary perturbations. *Nature* **313** 17–22.

Mikulic, D. G., Briggs, D. E. G., and Kluessendorf, J. 1985. A Silurian soft-bodied biota. *Science* **228** 715–717.

Palmer, A. R. 1983. The Decade of North American Geology 1983 Geologic Time Scale. *Geology* **11** 503–504.

Raup, D. M. 1984. Evolutionary radiations and extinctions. In: H. D. Hoffman and A. F. Trendall (eds.): *Patterns of change in Earth evolution, Dahlem Konferenzen 1984*, Springer-Verlag, Berlin, 5–14.

Raup, D. M. and Sepkoski, J. J. 1984. Periodicity of extinctions in the geologic past. *Proceedings of the National Academy of Science* **81** 801–805.

Ritter, S. M. 1985a. Evolutionary diversification of post-Early Permian crisis *Sweetognathus* in the central and western United States. *Geological Society of America Abstracts* **17** 2.

Ritter, S. M. 1985b. Ecophenotypic diversification of the *Sweetognathus* lineage: a biostratigraphic and biofacies tool for the lower Permian. In: R. J. Aldridge, R. L. Austin, and M. P. Smith (eds.): *Fourth European Conodont Symposium (ECOS IV), Nottingham 1985, Abstracts*, University of Southampton, 27.

Stenseth, N. C. and Maynard Smith, J. 1984. Coevolution in ecosystems: Red Queen evolution or stasis? *Evolution* **38** 870–877.

Sweet, W. C. 1973. Late Permian and Early Triassic conodont faunas, in the Permian and Triassic Systems and their mutual boundary. In: A. Logan and A. V. Hills (eds): *Canadian Society of Petroleum Geologists Memoir* **2** 630–646.

Van Valen, L. 1973. A new evolutionary law. *Evolutionary theory* **1** 1–30.

Wei, K-Y and Kennett, J. P. 1983. Nonconstant extinction rates of Neogenic planktonic foraminifera. *Nature* **305** 218–220.

# Taxonomic Index

# General Index